THIS BOOK BELONGS TO:

POTATO PRODUCTION, PROCESSING & TECHNOLOGY

By

Wilbur A. Gould, Ph.D

Emeritus Professor,
Food Processing and Technology,
The Ohio State University
and
Retired Executive Director,
Mid-America Food Processors Association
and
Consultant to the Food Industrues

Potato Production, Processing And Technology

A technical reference book and textbook for students of food technology, food plant managers, product research and development specialists, food brokers, technical salesmen, food equipment manufacturers, and food industry suppliers.

ISBN Numbers are as follows: 0-930027-30-2

Library of Congress Cataloging-in-Publication Data

Gould, Wilbur A., 1920-
Potato Production, Processing And Technology,
by Wilbur A. Gould.
p. cm.
Includes Index
ISBN 0-930027-30-2 (hc : alk. paper)
1. Potatoes. 2. Potato Products. I. Title. II. Title: Potato Production, Processing, and Technology
TP444.P6G68 1999
664',80521- -dc21 99-36722
CIP

While the recommendations in this publication are based on scientific studies and industry experience, references to basic principles, operating procedures and methods, types of instruments and equipment, and food formulas, are not to be construed as a guarantee that they are sufficient to prevent damage, spoilage, loss, accidents or injuries, resulting from use of this information. Furthermore, the study and use of this publication by any person or company is not to be considered as assurance that that person or company is proficient in the operations and procedures discussed in this publication. The use of the statements, recommendations, or suggestions contained, herein, is not to be considered as creating any responsibility for damage, spoilage, loss, accident or injury, resulting from such use.

A PUBLICATION OF

CTI PUBLICATIONS, INC.
2 Oakway Road, Timonium, Maryland 21093-4247USA
1.410.308.2080 • FAX 1.410.308.2079
email: sales@ctipubs.com • web: ctipubs.com

PREFACE

I was born in Colebrook, New Hampshire, the son of Alice and Elmer J. Gould. I was the fourth child in a family of six with four boys and two girls. I do not remember much about my Mother as she died when I was only three years old. My early childhood was not marked with many memories other than the fact that we worked hard on our dairy and potato farm.

We milked over 90 cows twice daily and I had my share to milk some 20 cows by hand. My one memory was when in elementary school in the Fall of the year of having to fill up a barrel (450 lbs.) of potatoes that my Dad and his helpers had dug during the day. In the early years I could only fill the pail as I was too short and the pail was too heavy to empty. (I remember going to Australia in 1984 and seeing small children hand picking up potatoes as I had to do before doing my other chores. This sure brought back some memories). The barrels were loaded onto sleds and horse drawn into the barn for storage over night before shipping to market. We grew the Green Mountain variety and had yields over 500 bushels to the acre. It was a good cash crop for my family.

My second work with potatoes was in 1938, my Freshman year as a student at the University of New Hampshire. I found a part time job working for Ford Prince and Paul Blood separating potatoes using salt solutions into different specific gravity lots. I did not know why we were separating them then, but I sure do today and it was a moment in my life that I will never forget.

My third introduction to potatoes was in 1942 as a USDA Food Inspector working in the Red River Valley inspecting dehydration factories and certifying lots of dried potatoes for acceptance by the military. I traveled from Minneapolis, to Osseo, Fargo, Grand Forks, East Grand Forks, Grafton, Bermiji, Duluth, Centeria, WI and back to Minneapolis nearly every week for some 30 months. I saw some real great efforts to produce good dried potatoes and other root crop products.

My next experience was while on the faculty at the Ohio State University starting in 1946 working with the late Dr. H. D. Brown on potato chip research. In the early 1950's I met Dr. Ora Smith and was invited to participate in the production and Technology Potato Chip schools as a speaker on Quality Control. These schools have been most helpful to me in developing information I consider very valuable to the growth of this great industry.

Over the years, I have been privileged to visit many potato processing plants in several states and, yes, in Germany, France, the Netherlands, England, Japan, and Australia. Every firm has been wonderful and they have given me a carte blanche opportunity to see and learn all I can about their operation. These plants have been my laboratory and the personnel have been most helpful in my education of and about this most important commodity.

The first honor one could hope to achieve in this industry was in 1990 when the Potato Association of America bestowed on me an Honorary Life Membership in the Association. This I cherish very much. In 1998 and a tremendous surprise was the Life Time Achievement Award bestowed on me at the Annual Snack Food Association Circle of Honor ceremony. This was done with great professionalism including a video of my life and some of my accomplishments.

Potatoes mean a whole lot to me. For nearly 80 years, starting as a child back in the depression when potatoes were our only food, potatoes have been part of my diet. Potatoes are still my favorite food and I still enjoy them for breakfast, lunch and dinner. I prefer them baked, boiled, chipped, canned, dried, and fried. My wife makes a great potato salad out of the canned product and we delight in using the canned potatoes for our shish kabobs. We have great recipes to use with our potato chips as snacks for entertainment. Potatoes are used as a focal point for most meals and we delight in serving them to our family and friends. We know they are good tasting and good for us.

— Wilbur A. Gould

ACKNOWLEDGEMENT

There are many, many people to thank for their wholehearted cooperation, their willingness to share, their helpful thoughts, their facilities, their equipment, their technologies, and their vision for this great commodity and industry. I apologize for not listing everyone, but the following are most important to me and do represent some great amount of effort over the years in helping me and my interest in and of potatoes. My most sincere thanks to each of you and the many that I have left off this roster of valuable people in my quest for help, knowledge, and information:

Larry Burch, Potato Chip and Snack Food Association
John Cady, National Food Processors Association
E. C. Wittmeyer, The Ohio State University
Don Giles of Heat and Control,
Barbara G. Weber of CSC Scientific
Ron Gould, General Mills
Winston Bash, The Ohio State University
Jack Corriere, Utz Foods,
Les Mapp, Mike-Sell's Potato Chip Company
Ora Smith, Cornell University,
Dick Chase, Michigan State University,
Heinz Fleshner, Bahlsen Snacks GmbH & Company
Wayne Pate, Golden Flake Snack Foods,
Victor Sabatino, Guy's Foods,
John Hayes, Curtice Burns Snack Group
Jim Herr, Herr Foods Incorporated
Kim Frank, Agtron, Incorporated
Joe Petrowski, Akzo Nobel Salt, Inc.
Joe Wallace, Vanmark Corporation
Traver Smith, Magnuson Engineers
Jim Shufelt, Snack Food Association

Also, I sincerely thank Jessie Gould for her reading of this book and constant guidance in making it come to fruition and to Randy Gerstmyer of CTI Publications for his understanding of this effort and total cooperation in publishing this book.

— Wilbur A. Gould

TABLE OF CONTENTS

CHAPTER 1

Introduction and History

HISTORY OF THE POTATO

The white potato is a native of South America and was first cultivated by South American Indians some 8,000 years ago. It has been the staple commodity of the inhabitants of Chili, Columbia, Peru, Ecuador, and Bolivia. In 1537 the first recorded contact of the white man with the potato was made by Gonzal Jiminez De Quesada in the village of Sorocota about 7 degrees latitude N. Costellanos stated that "the potato has floury roots of good flavor, a gift very acceptable to Indians and a dainty dish for Spaniards". The data in Figure 1.1 show significant happenings in the history of the potato.

The white or Irish potato is one of the world's most important food plants. It follows wheat, rice and corn as a major food crop. Since its introduction in the 16th century to Europe from South America its adaptability and capability to yield good crops have made it a staple food in nearly every civilized country (over 130 countries). The world potato crop has an annual production of about 300,000,000 metric tons.

The Indians were the first to preserve their potato crop. They would cut up the potatoes and place them on the ground to dry. At night time the potatoes would freeze and after thawing out in the morning the natives would stomp the potatoes causing the water to go free. By doing this night after night and day after day, the potatoes would become dry and in powder form and then ready for storage. This process is known today as freeze drying. The difference today is that the potato would be cooked first to inactivate the enzymes before attempting to freeze dry them.

The United States is far outranked by Russia, Poland, and

FIGURE 1.1 — Significant Dates in the History of the Potato

Date	Activity
1536	Potatoes were first encountered by Spanish Conquistadors in Andes
1553	First mention of potato in the literature by Cieca in Chronicles of Peru covering the travels of Conquistador Pedro Cieza de Leon and published in Seville, Spain
1586	Sir Walter Raleigh brought potatoes to England
1598	Gerard describes potato plant in Virginia
1601	Clusius published description of potatoes and called them *Papas Peruanorum*
1620	Bauhia names potato *Solanium Tuberosum*
1662	Royal Society recommending planting of potatoes to prevent famine
1710	Potatoes introduced from Ireland by a colony of Presbyterian Irish in Londonderry, New Hampshire
1731	Miller mentions two varieties of sorts, the red and the white
1769	Sweden noted potato culture in the literature with a royal edict to encourage their cultivation
1784	Kircheuff stated that potatoes are now grown in England by the rich
1796	Essex County grew some 1700 acres for London market
1814	The Society National d'Agriculture de France gathered and displayed a collection of 115 to 120 varieties of potatoes
1830	Potato was now more or less an object of field culture in every county in England showing some 1,605,730 CWT
1835	The United States began processing potato starch
1845	Irish potato famine caused by late blight fungus, *Phytophthora infestans,* with more than 1 million people starving to death
1853	Potato chip was created at Moon's Lake House Restaurant in Saratoga Springs, NY by Indian Chef named George Crum
1861	Luther Burbank developed the "Burbank Russet" potato variety
1892	Potatoes were exhibited at Horticulture show in Calcutta
1971	International Potato Center established in Lima, Peru

China in the production of potatoes. These three countries produce over 50% of the World production, whereas the U.S. produces about 7%. The utilization of the U.S. potato crop is shown in Figure 1.2.

The potato is one of the cheapest foods in the world and since it has no decided flavor, it is widely accepted. The average American now eats over 140 pounds of potatoes each year. Today, about 2/3rds of the harvested U.S. crop is processed into frozen French fries, potato chips, dried potato products, and canned potatoes.

History of Dried Potatoes

Drying is the oldest form of potato preservation. Chano is the oldest form of potato preservation. The Incas cultivated the potato some 3,000 years ago and they spread the potatoes on the ground to freeze overnight and thaw the next day. This freezing and thawing caused open cells as a pathway for the moisture to escape. The liquid was squeezed out by stomping on the potatoes and the potatoes were allowed to dry and be stored for year around use.

Dried potatoes have been a staple item for many years, particularly with great emphasis during World War I. By cooking potatoes and making a rice potato product before drying, they were good. Man did not know about enzymes until the middle 1930's, when workers at the University of California studied the effects of enzymes on foods for freezing. Their efforts were utilized during World War II when massive amounts of potatoes and other commodities were blanched before drying. I happened to be working in Minnesota and North Dakota at the time as a food inspector and learned the true meaning of blanching. Without blanching diced potatoes before drying they developed an off flavor in less than 3 months after drying. The flavor was 'haylike' and not acceptable. We learned to test for enzyme inactivity following blanching before drying and developed some very good dried products.

Workers at the United States Department of Agriculture Regional Laboratories were responsible for improving the quality of dried potatoes and the development of dried potato flour. They

FIGURE 1.2 — Potatoes: Utilization of Crops, United States, 1992–1995

Uses	Crop Year 1992	1993	1994	1995
	(1,000 cwt)			
Sales				
Table stock	127,215	123,802	133,989	123,989
Processing				
Chips and shoestrings	48,455	48,987	49,299	47,908
Dehydration	38,078	40,795	41,381	44,465
Frozen french fries	112,496	121,087	136,531	129,029
Other frozen products	23,016	25,190	26,362	26,573
Canned products	2,710	1,879	2,503	3,413
Other canned products (Hash, stews, soups)	2,557	2,458	3,006	2,492
Starch, flour and other	1,610	1,691	2,176	1,668
TOTAL	228,922	242,087	261,258	255,548
Other sales				
Livestock feed	3,928	2,498	4,147	3,224
Seed	23,529	24,223	24,616	25,570
Diversion	2,043	—	—	—
TOTAL	29,500	26,721	28,763	28,794
TOTAL SALES	385,637	392,610	424,010	408,331
Non-sales				
Seed used on farms where grown	4,746	4,808	4,732	4,787
Household use; used for feed on farms where grown	1,177	1,123	1,146	958
Shrinkage and loss	33,807	30,152	37,166	29,530
TOTAL NON-SALES	39,730	36,083	43,044	35,275
TOTAL PRODUCTION	425,367	428,693	467,054	443,606

— = Not applicable

SOURCE: *Potatoes*, National Agricultural Statistics Service, U.S. Department of Agriculture.

FIGURE 1.3 — Potatoes: Per Capita Use, 1970–1997[1]

Year	Total Fresh & Processed	Fresh	Processed				
			Total[2]	Frozen	Chips & Shoestrings	Dehy-drated	Canned[3]
			(Pounds fresh weight equivalent)				
1970	121.7	61.8	59.9	28.5	17.4	12.0	2.0
1971	117.8	56.1	61.7	30.1	17.2	12.3	2.1
1972	119.4	57.9	61.5	30.3	16.7	12.4	2.1
1973	118.2	52.4	65.8	34.2	16.3	13.1	2.2
1974	117.2	49.4	67.8	35.3	15.7	14.5	2.3
1975	121.9	52.6	69.3	37.1	15.5	14.7	2.0
1976	125.2	49.4	75.8	41.8	15.8	16.3	1.9
1977	122.1	50.1	72.0	42.2	16.2	11.4	2.2
1978	119.6	46.0	73.6	42.6	16.6	12.1	2.3
1979	117.8	49.3	68.5	38.5	16.7	11.2	2.1
1980	114.7	51.1	63.6	35.4	16.5	9.8	1.9
1981	116.5	45.8	70.7	41.5	16.6	10.8	1.8
1982	115.1	47.1	68.0	38.6	17.1	10.4	1.9
1983	118.7	49.8	68.9	39.2	17.8	10.0	1.9
1984	122.1	48.3	73.8	43.7	18.0	10.3	1.8
1985	122.4	46.3	76.1	45.4	17.6	11.2	1.9
1986	126.0	48.8	77.2	46.3	18.2	10.9	1.8
1987	126.0	47.9	78.1	47.9	17.6	10.8	1.8
1988	122.4	49.6	72.8	43.3	17.2	10.4	1.9
1989	127.1	50.0	77.1	46.8	17.5	10.8	2.0
1990	127.7	45.8	81.9	50.2	17.0	12.8	1.9
1991	130.4	46.4	84.0	51.3	17.3	13.7	1.7
1992	132.4	48.9	83.5	51.0	17.5	13.2	1.8
1993	136.9	49.7	87.2	54.5	17.6	13.4	1.7
1994	140.2	48.6	91.6	59.3	17.1	13.5	1.7
1995	137.9	50.7	87.2	55.3	16.9	13.0	2.0
1996p	142.3	49.2	93.1	59.0	17.0	15.0	2.1
1997f	143.3	52.3	91.0	57.3	16.8	14.9	2.0

[1] Calendar-year estimates.
[2] Excludes flour and starch.
[3] Includes potatoes canned in soups, stews, and other combinations.

p = preliminary; f = ERS forecast.

SOURCE: Economic Research Service, U.S. Department of Agriculture.

used drum dryers and developed flake drying of potatoes and potato granules. Flakes and granules are widely used today as mashed potatoes, an ingredient in some snack and baked products. Dried diced potato products are used in soups;, salad manufacture and hash browns. Some dried diced potatoes are ground and used in extruded products.

Today there are over 10,000,000 tons of potatoes dried annually in some 70 manufacturing plants with most of the factories located in Idaho using process grade Russet potatoes. Dried potato products are, also, a big business in Europe, particularly Germany and Great Britain.

History of Potato Chips

Potato chips had their beginning in 1853 in Saratoga, New York. It seems that a railroad magnate named Commodore Cornelius Vanderbilt was having his dinner at a fashionable hotel in Saratoga Springs, New York. The fried potatoes that were served him were too thick and he sent them back to the kitchen. Chef George Crum, an American Indian, was put out by the return of his potatoes and he decided to make them paper thin before he fried them. When presented back to the Commodore, he was pleasantly pleased and liked the texture and crunch of the new potatoes. The potatoes became a fad in the hotel and in the East but were primarily a restaurant item.

In 1895 William Tappenden of Cleveland, Ohio made chips for his restaurant crowd and the neighborhood stores. His business flourished from the kitchen stove to a converted barn at the rear of his home as the 1st potato chip factory. Soon other firms followed as shown in Figure 1.4.

The industry flourished and following World War I, there were several chip plants scattered around the U.S. Cleveland, Ohio was the center of much activity and in 1931 a group formed the Ohio Chip Association and in 1937 they expanded across the country and formed the National Potato Chip Institute, the forerunner to the Snack Food Association of today.

At the present time there are some 125+/– potato chip factories using nearly 50,000,000 CWT. of potatoes with potato chips per capita consumption at some 17 lbs.

FIGURE 1.4 — Early Chip Manufacturers

Year	Firm	Location
1895	Tappendens	Cleveland, OH
1908	Leominster Potato Chip Co. (Tri-Sum)	Leominster, MA
1910	Mike-Sell's Potato Chip Co.	Dayton, OH
1913	Dan-Dee Pretzel & Potato Chip Co.	Pittsburgh, PA
1920	Perfect Potato Chip Co.	Decatur, IL
1921	Magic City Food Co. (Golden Flake)	Birmingham, AL
1921	Wise Potato Chip Co.	Berwick, PA
1921	Utz Potato Chip Co.	Hanover, PA
1924	Moore's Potato Chip Co.	Bristol, VA
1926	Laura Scudder	Monterey Park, CA
1928	Jay's Foods	Chicago, IL

History of Frozen Potato Products

French fries were a product of World War II. Miles Willard states that they were made by Belgians speaking French and sold to the GI's. The potatoes were cut and par-fried in the morning then given a final fry on the street corner or in the bistro. The same product was made in the U.S. restaurants after the War.

The first commercial production of frozen potato products was started in Maine in 1947 by Baxter and Brothers. They were blanched, fried twice, then frozen. The product could be prepared by heating in the oven. The next step was to freeze the par fried strips and allow the user to give them a second frying and serve. Today they are partially dried after par frying and then frozen. The frozen French fry industry is an innovative industry with additives to the fries to help their texture. The cuts are done with water knives leaving less waste and more uniformity to the long cuts. They were one of the 1st to use electric eye scanning to remove defective slices and this industry has developed technology to utilize all of the potato with many innovative products. Miles Willard is one of the real innovators.

FIGURE 1.5 — Number of Potato Chip Plants and Quantity Used for Chips and Shoestring Potatoes by Area and State, 1992-1995 Crops

Area & State	1992 Plants (no.)	1992 Quantity (1000cwt)	1993 Plants (no.)	1993 Quantity (1000cwt)	1994* Plants (no.)	1994* Quantity (1000cwt)	1995** Plants (no.)	1995** Quantity (1000cwt)
New England CT, ME MA, NH RI, VT	10	2,859	10	2,954	9	2,981	8	2,932
Eastern DE, DC MD, NJ NY, PA, VA	31	10,548	31	10,932	30	9,970	28	8,992
North Central MI, OH WV	22	4,296	21	4,107	20	4,297	20	4,513
Mid-Central KS, MO, NE	8	2,304	7	2,168	7	2,309	7	2,253
Midwest IL, IN, IA MN, ND SD, WI	22	8,269	20	7,718	17	7,278	17	7,863
Southeast AL, FL, GA KY, LA, MS NC, TN	20	8,187	20	9,031	18	9,015	18	9,470
Southwest AR, OK, TX	6	3,626	6	3,667	6	4,089	6	3,987
Rocky Mtns. CO, ID, MT NM, UT, WY	9	2,113	7	1,865	4	2,484	4	1,981
West Coast AZ, CA, NV OR, WA, HI	19	5,562	19	5,912	20	6,260	18	5,367
Total	147	47,764	141	48,354	131	48,683	126	47,358
Shoestrings	–	691	–	633	–	616	–	550
Total used for chips & shoestring#	–	48,455	–	48,987	–	49,299	–	47,908

– = Not available. *Revised. **Preliminary. #Totals may not sum due to rounding.

Source: Potatoes, National Agricultural Statistics Service, U.S. Dept. of Agriculture.

Tuber Composition

The composition and amount of stored nutrients depend upon the growing season temperature and moisture, the soil and nutrient elements in the soil, and the genetic makeup of the cultivar or variety. The composition is of particular importance and concern to the potato manufacturers. The approximate analysis of white potatoes is given below. Approximately 80% by weight of potato carbohydrate is starch and is composed of amylopectin (75-95%) and amylose (21-25%). Potato protein is made up of all essential amino acids. Potatoes are, also, an excellent source of Vitamin C and other nutrients. The wide range of nutrients indicate many variables may affect the composition and, in part, explain why tubers differ widely. Without question the variety or cultivar is the number one cause of composition difference.

FIGURE 1.6 — Potato Composition

Constituent	Average	Range in Percent
Water	77.5	63.2 - 86.9
Total Solids	22.5	13.1 - 36.8
Carbohydrate	19.4	13.3 - 30.5
Protien	2	0.7 - 4.6
Crude Fiber	0.6	0.14 - 3.48
Ash	1.0	0.4 - 1.9
Vitamin C	12	5.0-20 mg./100 gm

FIGURE 1.7 — Potato Chips Composition

Constituent	Range in Percent
Water	up to 2.00%
Potato Solids	60–70%
Fat	25–35%
Salt/Seasoning	1–3%

CHAPTER 2

Potato Production

Potatoes grow best in comparatively cool and moist climate and in lighter soils. They are grown extensively in Northern Europe, Ireland, Northern United States, and Canada. It is reported that it was first produced in Londonderry, New Hampshire in 1719 from stock brought from Ireland. They are presently grown in nearly every state in the United States with concentration in Maine, New York, Michigan, Wisconsin, Minnesota, North Dakota, Idaho, Washington, Oregon and California. Potatoes are grown on some 1,400,000 acres in the U.S. About 15% of the U.S. crop is produced in the southern states, particularly Florida. (See Figure 2.2) Yields vary upwards to 1,000 CWT/acre, but average around 350 CWT/acre.

Anatomy of the Potato

The potato is a member of the nightshade family, *Solanaceae*. The potato is an annual, herbaceous dicotyledon as far as its vegetative and flowering habits are concerned, but it may be regarded as a perennial as far as its capacity for reproduction by means of the tubers. Thus, it is generally propagated asexually by cut pieces of tubers. At the time of flowering, tuber formation is usually initiated and involves the enlargement of the tip of the rhizome. The rhizomes arise from the underground portion of the stem. Many factors may affect the formation and number of tubers, such as temperature, moisture supply, length of light period, and formation of carbohydrates.

The tuber which is the edible portion of the plant, consists of a thickened, underground stem. Tubers may be dug before they are fully mature to meet the demands of the user. Usually the immature potato has skin or periderm that is fluffy with curled fragments. Immature tubers should not be stored as they may break down quite quickly if not used promptly.

Potatoes are grown from seed, that is, cut pieces of a potato with an eye in each piece The seed piece should be $1\frac{1}{2}$ to 2 ounces for best yield. The seed pieces should be uniform in size and weight for uniform stand and yield per hill. The seed piece should be treated to prevent disease and, of course, kept clean and only planted with clean planters to prevent disease inititation. The seed piece should be properly spaced in the row and planted 3 to 4 inches deep. Row spacing varies with soil type, but 10 to 12 inches between pieces is a good guide. This means one would need 100 to 120 seed pieces per 100 foot row.

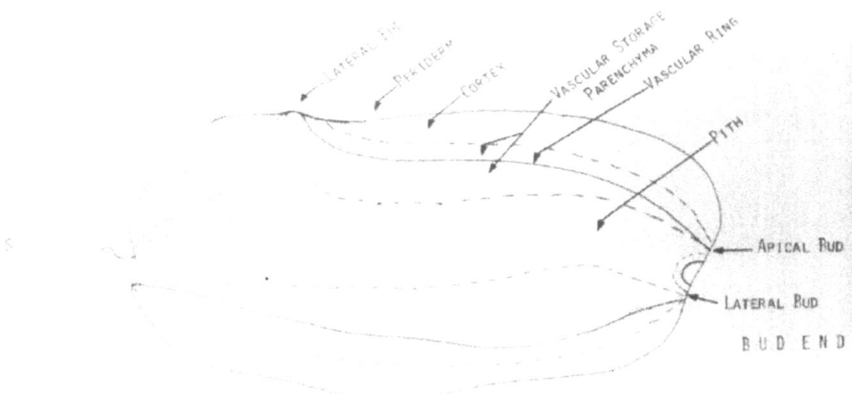

FIGURE 2.1 — Anatomy Of A Tuber

Culture

Since potatoes are grown in every state and in many parts of the World, the culture is somewhat different for each potato producing area. There are many authorities on culture, and no attempt will be given here to describe any of the areas other than to point up some significant facts.

Fertilization is an exacting practice for best results in crop yield and potato quality. One should rely on extensive soil testing and make application of fertilzers accordingly. County agents and soil testing laboratories are most important sources of information.

FIGURE 2.2 — U.S. Potatoes: Area Harvested, Yield and Production, 1990–94 Average and 1995, 1996*

State	Harvested area 1990-94 aver.	Harvested area 1995	Harvested area 1996	Average yield per acre '90-94 aver.	Average yield per acre 1995	Average yield per acre 1996	Production '90-94 aver.	Production 1995	Production 1996
	—1,000 acres—			—CWT—			—1,000 CWT—		
Alabama	10.3	9.2	8.6	148	167	152	1.523	1.539	1.309
Arizona	6.2	6.5	9.0	277	270	275	1,704	1,755	2,475
California	44.9	41.3	43.1	363	354	363	16,287	14,620	15,651
Colorado	76.6	85.9	87.5	343	309	364	26,301	26,584	31,890
Delaware	6.3	5.9	5.9	199	250	250	1,261	1,475	1,475
Florida	43.2	42.9	44.3	207	210	198	8,963	9,003	8,788
Idaho	394.0	398.0	408.0	321	333	343	126,658	132,657	139,960
Illinois	4.2	5.5	6.9	273	270	260	1,157	1,485	1,794
Indiana	4.1	4.6	5.2	240	260	250	989	1,196	1,300
Iowa	1.2	1.6	1.5	180	150	200	216	240	300
Maine	77.8	78.0	77.0	260	220	280	20,251	17,160	21,560
Maryland	2.0	1.5	1.9	153	240	200	303	360	380
Mass.	2.9	3.3	2.7	235	260	260	690	858	702
Michigan	48.7	54.5	46.0	275	300	300	13,414	16,350	13,800
Minnesota	72.7	77.0	82.0	244	270	300	17,725	20,790	24,600
Missouri	6.6	6.9	7.0	222	230	230	1,457	1,587	1,610
Montana	9.1	9.8	10.2	299	300	315	2,729	2,940	3,213
Nebraska	12.3	15.9	18.0	308	310	335	3,788	4,934	6,025
Nevada	7.6	7.6	8.0	357	365	390	2,700	2,774	3,120
New Jersey	3.6	2.6	2.5	213	270	240	765	702	600
New Mexico	9.6	10.5	10.6	382	356	371	3,669	3,738	3,928
New York	28.4	28.5	28.5	269	270	280	7,623	7,695	7,980
N. Carolina	18.3	17.8	17.4	180	178	169	3,292	3,177	2,943
N. Dakota	134.4	121.0	131.0	184	210	220	24,737	25,410	28,820
Ohio	6.5	5.4	5.1	222	260	250	1,453	1,404	1,275
Oregon	50.6	53.2	64.0	463	466	499	23,462	24,788	31,925
Penn.	19.9	17.0	16.5	223	240	255	4,444	4,080	4,208
Rhode Isl.	1.2	0.9	0.8	231	270	300	272	243	240
S. Dakota	6.6	5.2	5.7	240	190	280	1,572	988	1,596
Texas	13.8	12.0	16.0	210	214	212	2,912	2,570	3,385
Utah	6.1	5.1	4.2	268	240	280	1,629	1,224	1,176
Virginia	10.7	8.5	8.3	161	240	175	1,726	2,040	1,453
Wash.	141.4	147.0	161.0	552	550	560	78,027	80,850	90,160
Wisconsin	68.1	80.0	81.0	352	325	390	23,968	26,000	31,590
Wyoming	1.9	1.5	0.8	270	260	280	502	390	224
U.S.	1351.9	1372.1	1426.2	317	323	345	428,169	443,606	491,455

*1996 data are preliminary. Data are for all four seasons.

SOURCE: *Crop Production and Potatoes*, National Agricultural Statistics Service, U.S. Department of Agriculture.

Nitrogen is most important for the production of healthy plants and it must be available throughout the growth of the potato. However, an over supply of nitrogen will result in smaller tubers and will delay harvest. Excess nitrogen may be responsible for growth cracks in the tubers and it may lower specific gravity.

Phosphorus is needed to stimulate the seed piece and the proper development of roots and stem growth, the initiation of blooming and phosphorus aids in seed formation.

Potassium promotes root growth maintaining the plant in upright position, reduces wilting and is essential in the translocation of sugars.

Other essential nutrients for plant growth are sulfur, calcium, magnesium and trace elements like boron, copper, iron, manganese and zinc.

FIGURE 2.3 — Cross Irrigation System

The amounts of fertilizers are determined by soil tests and for the production of good quality crops the results of the soil test must be followed. The results of the soil test are only as good as the sample, thus, proper sampling of the fields is most important.

In general one needs nitrogen to obtain satisfactory yields. It has been reported that rates vary from 200 lbs. N to as high as 600 lbs. N, with yield ranging from 200 to 500 CWT per acre. Phosphorus (P) and Potassium (K) will vary too with soil tests,but one can expect to add P in amounts up to 100 lbs. per acre and K in amounts up to 200 lbs. per acre.

Petiole and leaf analysis is a tool to help develop the exact fertilizer usage during the growing season. It is a relative new tool to help the grower know when to add more fertilizer, particularly nitrogen during the growth of the crop.

Weed control, insect control, and disease control are specific subjects for given areas of the country depending on the problems at hand. The processor is concerned with the correct useage of control materials, their amounts and, of course, the timing of the applications. Residues can be costly and may mean the total destruction of the crop.

Moisture may need to be applied to a potato crop during the growing cycle. The amount and timeliness of irrigation is most important to produce quality potatoes.

FIGURE 2.4 — Potato Field In Bloom

Tuber Maturity

During the growth of potatoes, carbon dioxide from the atmosphere and water from the soil combine in the green plant when exposed to sunlight to form simple sugars (reducing sugars, that is, glucose and fructose) in a process called photosynthesis. The simple sugars that are formed in the plant are synthesized to form sucrose. Sucrose is then translocated from the above ground stem to the underground stem, that is, the rhizome or tuber. Here the sucrose is further synthesized to form starch which is then stored in the tuber and will continue to accumulate until the plant stops manufacturing sugars. The whole process of reducing sugars to sucrose to starch may be reversed or hydrolyzed during respiration. Respiration may be excessive under stress conditions, that is, lack of water, too high temperatures, too low a temperature, excess water, fertilizer imbalance, early vine kill which may be due insect damage, disease or drought. The build up of total solids (starch and other carbohydrates, amino acids and proteins, fats, minerals and vitamins) is determined by the growth and health of the potato plant.

Mature tubers are those tubers that come from fields where the vines have died or have been killed with the tubers left in the ground for upwards of two weeks before digging. Vine killing may be accomplished by roto beating the vines or by use of specific vine killing chemicals. Of course, vines may be killed naturally, that is by frost. Vine killing tends to toughen up the potato skins on the tubers and may prevent some skinnning and bruising. The killing of the vines, also, aids in the harvesting of the potatoes.

Potatoes to be stored should be mature. Maturity has been described as skin (periderm) set and lack of further growth of the tubers. Maturity can be measured by determining the amount of sucrose in the tuber. Sucrose or Chemical Maturity Monitoring (CMM) should be 1% or less if one wishes to store tubers for the chip market. As shown in Figure 2.5 sucrose should be monitored during growth and storage. Harvesting should take place when the sucrose is below 1.5% sucrose. During the storage life of the potatoes, it should remain at this level or lower. When the sucrose level rises in storage the potatoes must be used or dark chips will result. In most cases sucrose will be too high for any amount of reconditioning to lower the glucose level to 0.15%.

FIGURE 2.5 — Typical Changes In Sucrose Concentration

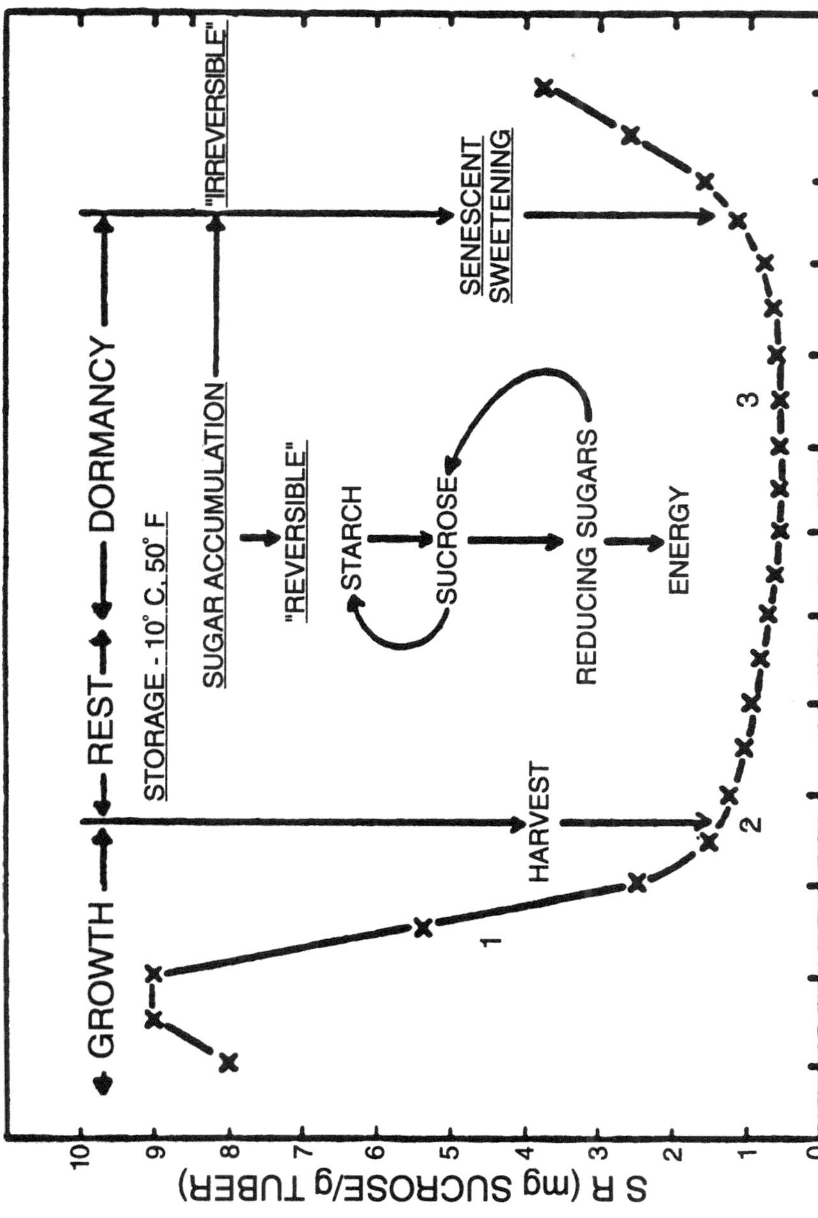

If the glucose (reducing sugar) level is too high (>0.20%) these sugars will react with the naturally present amino acids (Mallaird reaction) and cause dark colored chips. Thus, when manufacturing chips, the only sugar test that need be made is the reducing sugar test whether using the Snack Food Association (SFA) sugar test tapes or using the YSI analytical test. The YSI method adapted at The Ohio State University is an easy and very adequate method for evaluating the potatoes for chip making, that is, glucose testing or for storage evaluation, which is, sucrose measurement.

Reducing sugars can be determined with the SFA glucose tape as follows:

1. Select 5 or more tubers at random from the lot of tubers in question and cut one tuber at a time in half lengthwise. (The knife should be cleaned prior to cutting each tuber).
2. Withdraw enough SFA Chip Color test tape or ribbon from the dispenser to extend the entire length of each cut tuber. Handle the tapes from the end only and apply to the cut surface of each cut tuber. Press the cut surfaces of each tuber together so that the tape is moistened with the natural tuber juices. Immediately separate the two halves and remove the test tape.
3. Wait one minute and then match the color of the tuber with the color chart on the ribbon dispenser. If the ribbon remains yellow, it indicates that no glucose (reducing sugar) is present and the potatoes should fry to an acceptable color. If the ribbon turns green and is darker than 1/10% from the color chart, the chips will probably fry dark.

   Note: If the ribbon at the stem end of the potato is green, the chips made from the tuber will have a dark end to the chip as excess reducing sugars have not left the tubers.
4. The test tubers should be discarded and not used.

The YSI (Yellow Springs Instrument from Yellow Springs, OH and available in most laboratory supply houses) Glucose and Sucrose method for evaluating tubers for chipability from the field or for storage of tubers is a relatively simple method and quite accurate method to evaluate sugars in potatoes.

1. Select 100 grams of peeled sliced potatoes from 10 or more tubers or remove 10 plugs (1/2 inch diameter) from 10 tubers equaling 100 grams and place in a blender.
2. Add 150 ml of distilled water to the blender bowl.
3. Blend the tubers and water for 45 seconds.
4. Filter the mash (#3 above) through a #588 S&S filter paper.
5. Remove a 25 micro liter aliquot of the filtrate and inject with a Springerpet (YSI 2361) into the previously standardized YSI Model Analyzer.
6. The reading on the instrument is the free glucose designated as Glucose 1.
7. Transfer 3 ml of the filtrate (#4 above) into a 56 ml polystyrene micro beaker (YSI 2362).
8. Add 2 ml invertase enzyme (Sigma 1-4505) to the filtrate in the micro beaker and thoroughly mix.
9. Hold the mixture in #8 above at room temperature for 30 minutes to complete the hydrolysis of the sucrose to glucose.
10. Remove a 25 micro liter aliquot of the mixture in #9 above and inject with a Springerpet (YSI 2361) into the YSI Analyzer.
11. The reading obtained is the total free glucose as in #6 above and the glucose produced by sucrose hydrolysis as in #8 above is designated glucose 2.
12. The glucose and sucrose levels are indicated as follows:

    Glucose (mg/g) = mg/dl x 150 ml/100 g

    Sucrose (mg/g) = Glucose 2-Glucose 1 x 1.91 x 150 ml/g

Harvesting of Potatoes

Harvesting of potatoes is a mechanical process where the potatoes are lifted from the field with a mechanical digger that takes potatoes, soil, vines and other extraneous material into the throat of the harvester. The tubers are then separated by shaking the tubers from the vines and separating the soil from the tubers as they are conveyed through the machine. The vines and soil drop back on the ground and the tubers are conveyed to bins or trucks for further conveying them to the storage facility.

Bruising is without doubt the second most important factor when working with potatoes for the chip market. Bruises developed during harvesting, handling and storage of tubers create a great loss of raw product. There are 4 factors that are believed significant in influencing bruises in potatoes, that is, soil condition at time of harvest, tuber condition, temperature, and the operation of the harvester.

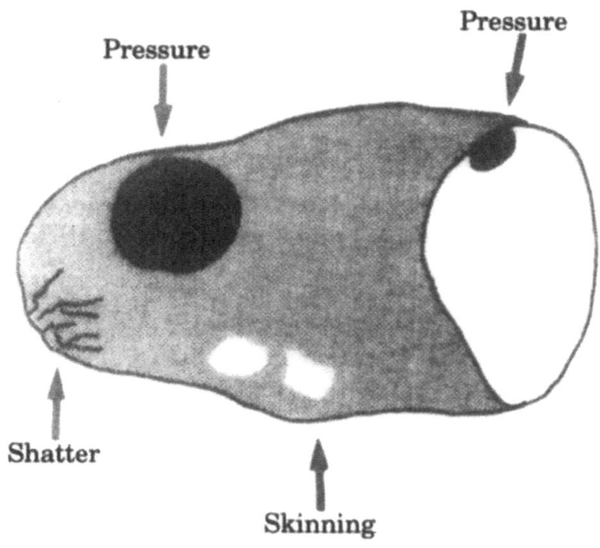

FIGURE 2.6 — Types Of Potato Bruises

Soils that are heavy and wet are difficult to separate the tubers from the soil. Moisture in the soil should be between 60 and 80% of the field capacity for loam and sandy soils. Heavy dry soils form clods which increase the damage to tubers while inside the harvester. Of course, soils with stones present will also bruise the tubers during harvesting.

Tuber condition is of direct concern and may involve delaying harvest after vine killing to allow the tubers to better set their skin. Turgid tubers (firm) tend to shatter bruise while flaccid tubers (soft) tend to develop black spot. Large tubers bruise more easily than small tubers. Further, high specific gravity tubers appear to bruise more easily than those with lower specific gravities.

Tuber bruising is one of the most serious problems facing the producer and the processor. Skinning is probably one of the easiest type of bruise to eliminate. Tubers should only be handled when they are completely mature. Top killers should be used 2 weeks prior to harvest to allow the tuber to mature.

To prevent shatter bruise (cracks or splits on the tuber surface that penetrates the flesh) and black spot (dark semispherical spot in the flesh beneath the hard tuber surface) development, tubers should never be handled when the temperature is below 45 degree F. (7.2 C.). Preferably, tubers should only be handled when the temperature is above 55 degree F. (12.7 C.).

Another important type of bruise is the so-called pressure bruise which occurs in storage and is most severe when the humidity is low, that is, below 90%. These tubers develop softened, flattened or indented areas as a result of continuous pressure. They may discolor the flesh of the tuber which shows up after processing as gray areas.

Bruises can be detected by using the catechol test. Catechol is a polyphenol compound that reacts with the naturally present enzymes in the tuber and turns the damaged flesh dark. The test is of value to the harvester operator and the handler to know if damage is being done with their system of harvesting and handling. Bruised potatoes are not desired by the industry and it behooves operators to learn to do things correctly to eliminate this severe appearance problem and loss to the producer and the processor.

The movement of the harvester chains, that is, the speed of the chain, the prevention of tuber roll back, and the elimination of tuber drops are all factors of the operation of the harvester that may affect bruising of tubers during digging. Harvester speed and adjustments must be made depending upon the local conditions. Excessive drops and roll down in handling the tubers lead to skinning and ultimate bruising of the tubers. Rubberized padding is a must to prevent bruising.

Storage of the Potato

The goal of potato storage is (1) to retain the natural water in the tubers, (2) to hold the respiration rate to a minimum, (3) to hold reducing sugars to near zero, and (4) to maintain the external appearance of the tubers.

Freshly dug potatoes must be properly cured or suberized, that is, healed from digger and/or other mechanical damage before low temperature storage. The humidity should be high, that is, 95% in a well ventilated area to allow suberin formation. A layer of suberin or periderm can develop in less than 48 hours with 3 layers in six to seven days. It is essential that the periderm be formed to prevent rot or microorganism penetration. Temperatures during the suberization process should be maintained at 60°F (15.6°C).

After the wounds are healed, the temperature may be gradually lowered to the desired storage temperature. This is determined by the variety of tubers in use and the length of the storage period. The temperature should be low enough to allow the tuber to go into its rest stage to prevent weight loss and excessive peel loss at time of use. Tubers can be stored for up to 10 months if storage conditions are right, that is, temperature and relative humidity are provided, regulated, and maintained through the use of correct airflow to provide proper ventilation and complete air distribution throughout the storage pile.

Some suggested storage parameters are:

Temperature @ 50°F (10°C) to as low as 38°F (3.8°C) depending on varieties and ultimate use.

Relative Humidity @ 90 to 95%

Air Flow Rate @ 0.04 cubic meters/kg-hr.

According to Ora Smith, "Potatoes usually undergo a rest period of several months during which there is little or no sprout growth regardless of environmental conditions. Following the termination of the rest period, sprout growth occurs at temperatures above 41°F (5°C). If storage conditions could be maintained below 41°F (5°C) very little, if any, sprout growth would occur". If sprouting does occur there is an increase in loss of moisture and pronounced shriveling of the tubers causing difficulty in peeling of the tubers.

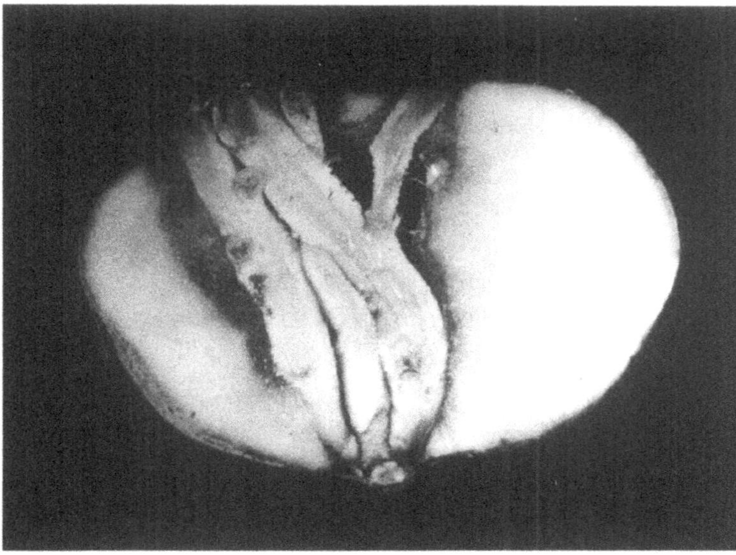

FIGURE 2.7 — Internal Sprouting Of Potato

At the present time there are several measures to control sprouting of tubers. As stated above, temperatures below 41°F (5°C) will control sprouting of most cultivars, but sugars may accumulate. Low temperatures require the use of cultivars that are fully mature and cultivars that do not accumulate reducing sugars. A major effort is being set forth today to develop cultivars that are known as "cold chippers", that is, potato cultivars that can be stored as low as 38°F (3.3°C). These cultivars, hopefully, will have a long dormant cycle.

As an alternative to low temperature storage, tubers may be treated with sprout inhibitors. Several chemicals have been used as sprout inhibitors. The most common chemical is Chloro IPC (isoprophyl N-chlorophenyl carbamate) for treating only suberized tubers for long term storage. This compound may be applied 2 weeks after suberization as a liquid, a powder or by fogging (gas). The gas must cover all the eyes of the tuber. The tolerance level is 50 ppm in the tuber. Other chemicals that have been used are Nonanol, Maleic Hydrazide (MH-30) or Tecnazene (Fusarex).

FIGURE 2.8 — Potatoes Hauled In Truck For Unloading

FIGURE 2.9 — Truck Dumper System

MH30 is the most commonly used chemical of this group and it is applied in the field on the foliage during the growing season after blossom drop, but at least 2 to 3 weeks before vine killing. It is applied at the rate of 1 gallon/acre with sufficient water to get complete coverage. The tolerance level is 50 ppm in the raw potato and 160 ppm in potato chips.

Fusarex is applied as a dust on tubers moving into the storage at the rate of one pound of the 6% formulation per 600 lbs. of potatoes.

A third alternative for tuber storage is the use of controlled atmospheric conditions (CA storage). CA storage implies control of oxygen levels. It means mechanical cooling and gas tight buildings. The object is to put the tubers to sleep, that is, cut off the oxygen levels and reduce the respiration levels or slow down the tubers release of carbon dioxide. Obviously, the gases must be carefully controlled while the tubers are dormant.

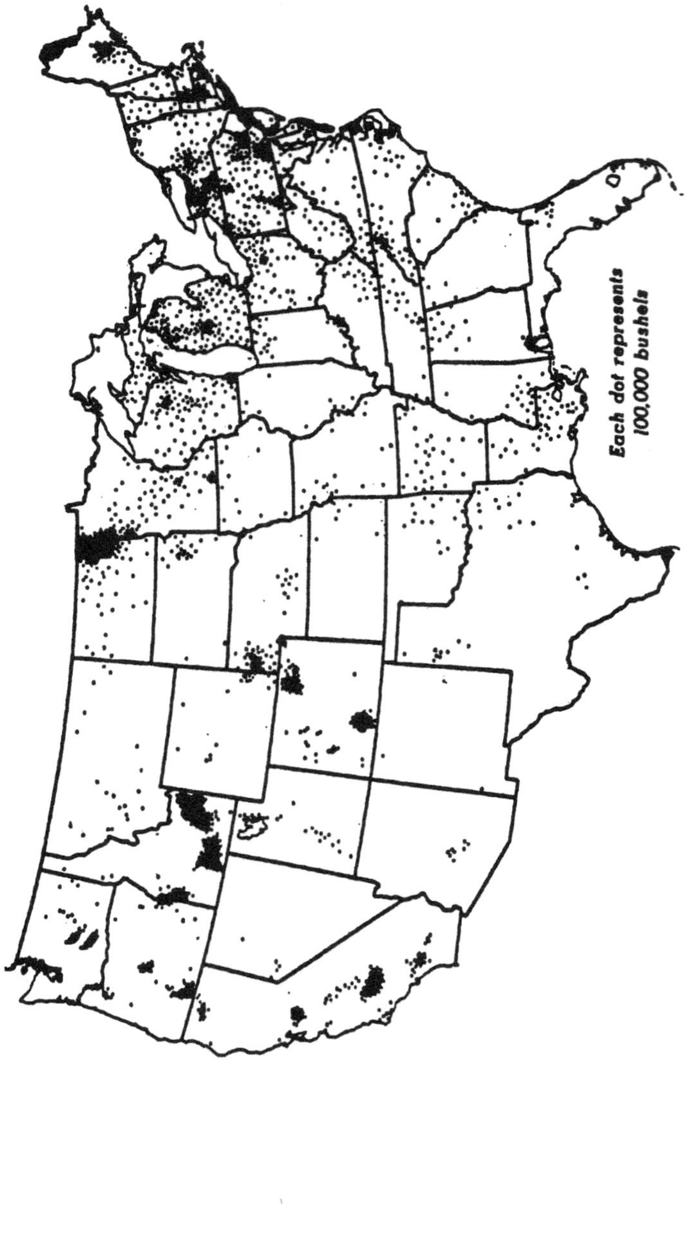

FIGURE 2.10 — Locations of Potato Growing in The U.S.
(Courtesy U.S. Department of Agriculture

FIGURE 2.11 — Principal Producing Areas of Irish Potatoes, by Seasons (Courtesy of U.S. Department of Agriculture

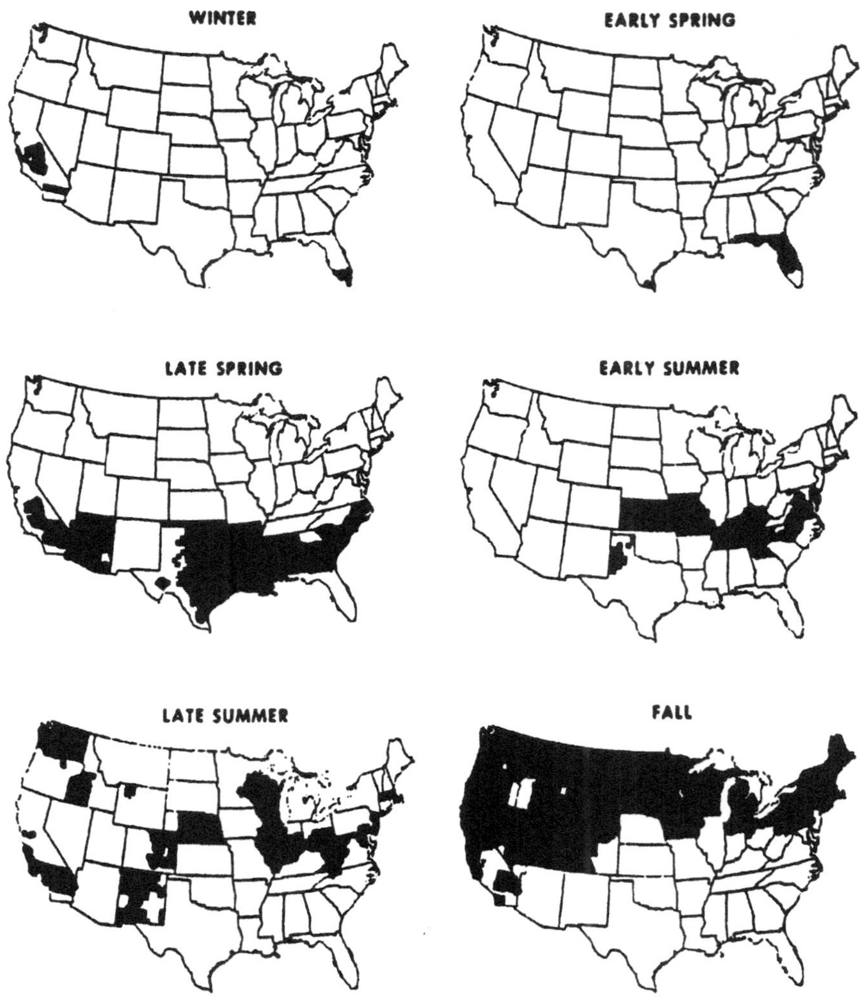

CHAPTER 3

Potato Diseases

The material used in this section of this book is taken principally from W. J Hooker, *Compendium of Potato Diseases* (APS Press) and G. B. Ramsey et al *Market Diseases of Potatoes* (USDA Misc. Publ. 98) with a discussion of Late Blight from Randy Rowe (presentation at NPC/SFA Chipping Potato Seminar at SC in 1977)

"A potato disease is an interaction between a host (the potato) and a pathogen (bacterium, fungus, virus, mycoplasma, nematode, or adverse environment) that impairs productivity or usefulness of the crop".

Only a few of the various diseases will be presented below:

BLACKHEART—Results from inadequate oxygen supply for respiration of internal tuber tissue. The symptoms consist of black to blueblack discoloration in irregular patterns in the central portion of the tuber. Control is to not expose tubers to high temperatures nor to low storage temperatures and if in closed bins provide forced aeration of the potatoes.

HOLLOW HEART—Associated with excessively rapid tuber enlargement. Incidence is associated with large tubers.

BLACKSPOT—Always caused by bruising injury, either from impact during harvest, handling, and grading, or from pressure during storage. The symptoms show up as bluegray to black discolored areas developed just beneath the skin. Bruising injury initiates a series of biochemical oxidation's in damaged cells. Phenyl substrates such as tryosine are oxidized to conjugated quionones by polyphenol oxidases. The quinones polymerize to produce black pigments.

BLACKLEG, BACTERIAL SOFT ROT—Caused by *Erwinia carotovora* and may be found any where potatoes are grown. Stems, petioles, leaves and tubers may be infected. Control is by using clean seed and avoidance of moisture during harvest and handling.

EARLY BLIGHT—Disease is found worldwide and is caused by *alternaira solani.* Clean seed must be used and fields must be kept clean. Use of appropriate fungicides is a requirement.

SILVER SCURF "A"—Fungus growth on tubers caused by *Helminthosporium solani.* Control by using disease free seed, harvest when tubers are mature, and ventilate storage properly.

FUSARIUM DRY ROT—Found on potatoes worldwide. Affects tubers in storage. Provide high humidity and good ventilation in storages to facilitate wound healing and provide aeration during storage.

FUSARIUM WILT—Widespread and most severe when potatoes are grown at relatively high temperatures or when seasons are hot and dry. Use clean seed and avoid contamination by inoculum transfer through infested soil or diseased tubers.

VERTICILLUM WILT—Use disease free seed and resistant varieties. Also, practice proper rotation and control weed suspects.

LATE BLIGHT—The single most important disease of the potato worldwide. It was the cause of the Irish famine in 1840's and the causal organism is *Phytophthora infestans.* Late Blight symptoms appear as small, dark spots on the upper surfaces of the leaves (See Disease Cycle in Figures 3.1 through 3.4). These enlarge into large, irregular, brown-black lesions with light gray-green to light brown margins. A pure white mold growth is often visible on the underside of infested leaves if the conditions are moist. Eventually these lesions enlarge into purplish black areas covering large portions of infested leaves which eventually die and fall to the ground. The fungus may infest the stem and may be up to 3 or 4 inches in length and nearly encircle the stem. The late blight may also infest the tuber with infected tubers appearing somewhat darker brown than healthy tissues and are usually sunken and shrivel. When cut, the infected internal tissues appear copper colored and granular in texture. One must

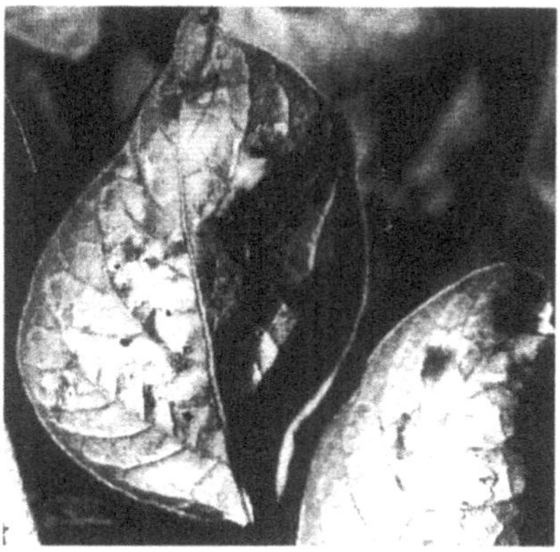

FIGURE 3.1 — Large, irregular late blight lesions on leaves of potato are blackish-brown and usually have light brown or light green margins. Lesions on tomato leaves are similar.

FIGURE 3.2 — Underside of the late blight lesion shows cottony, white mold growth of the late blight fungus. This may be absent in dry weather.

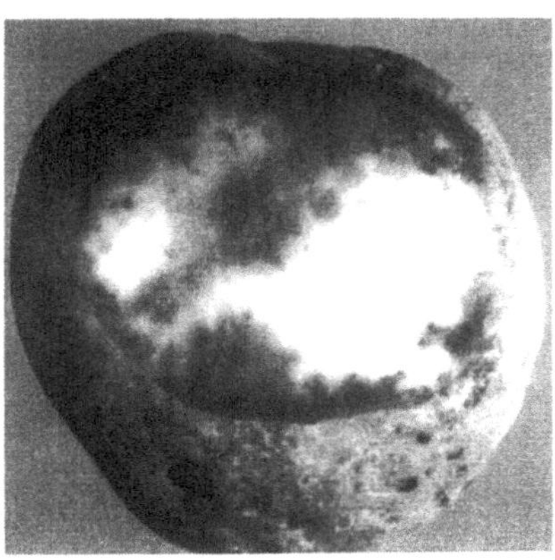

FIGURE 3.3 — Potato tuber tissues with late blight are firm, and, when cut, appear tan to coppery-brown with a granular texture.

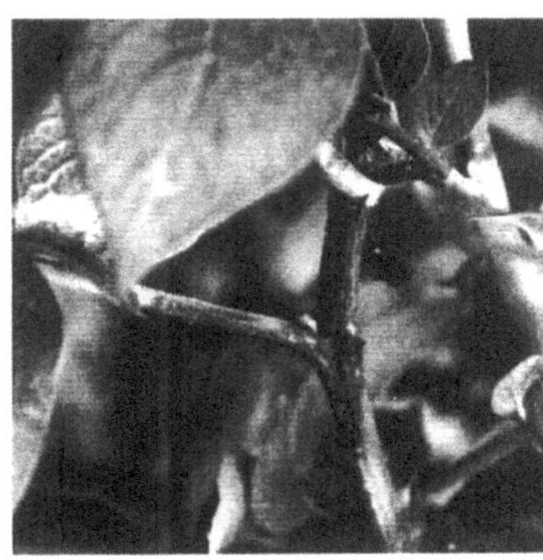

FIGURE 3.4 — Purplish-black late blight lesions can appear anywhere on potato stems.

understand the biology of the late blight for control. Control is by elimination of sources of disease carryover, such as cull piles and volunteer plants. One should use certified disease-free seed, and follow the correct use of fungicides and good cultural practices. In addition, it is important to have adequate ventilation of storage's.

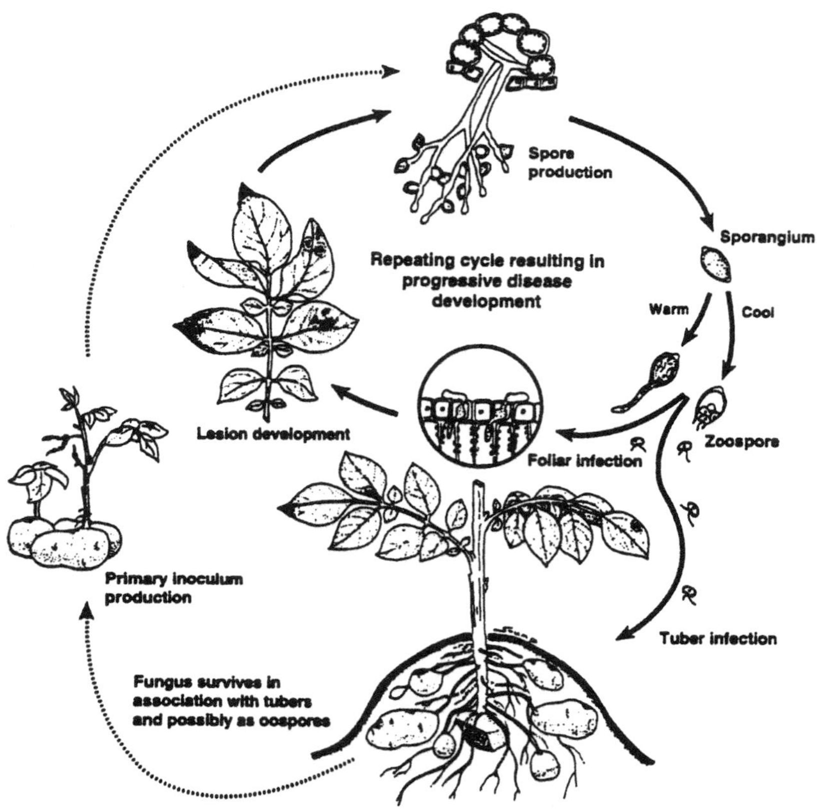

FIGURE 3.5 — Late blight disease cycle. The fungus survives in infected tubers and volunteer plants. Spores are only produced on the surface of living host plant tissue. Disease progression is influenced by many factors, including host susceptibility, fungicide use and environmental conditions. Cool wet conditions favor disease progression. Simultaneous plant infection by the A1 and A2 mating types may result in sexual spore (oospore) production and soil survival. (Prepared by G.D. Franc and W.L. Stump.)

FIGURE 3.6 — Chipping Potato Disease & Defect Chart (Courtesy Snack Food Association)

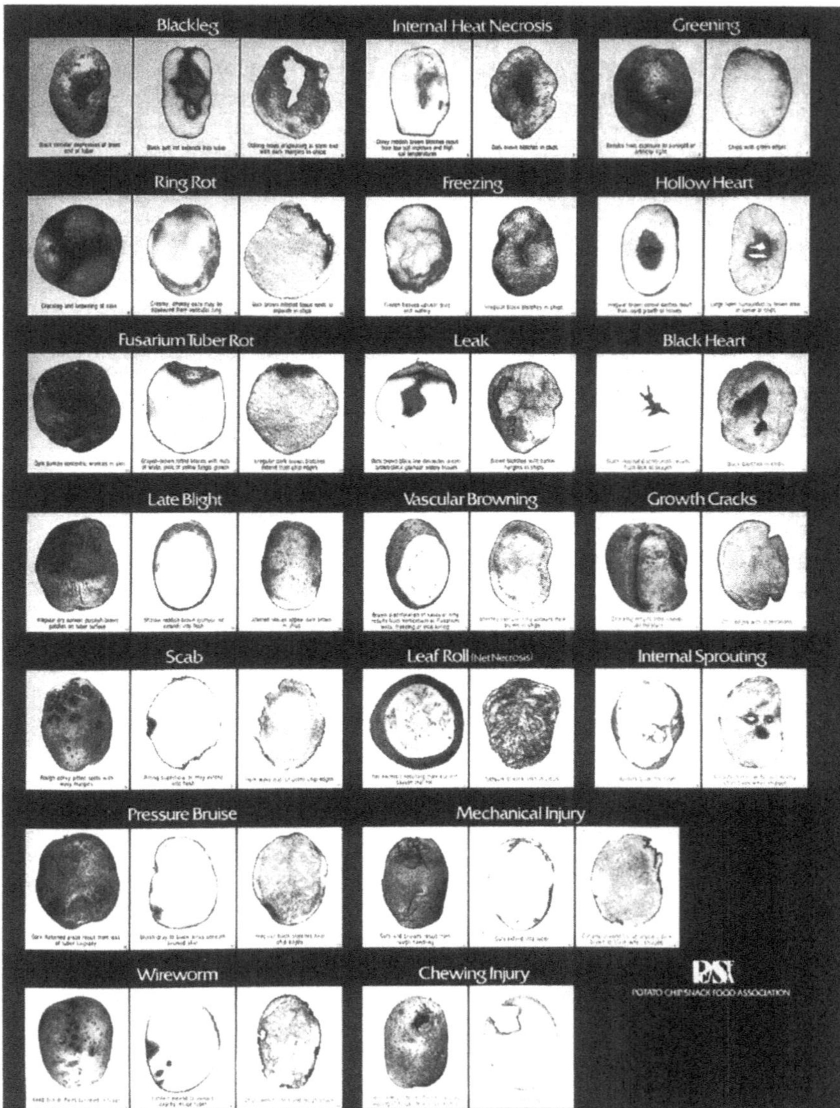

There are many viruses and insects, such as, nematodes and aphids that cause serious problems in given potato production areas. The references cited above and most State Universities can provide information on direct control of these and other potatoes diseases. Many of the above diseases and disorders present serious problems to the user of potatoes and it behooves the potato producer and/or handler to use every precaution possible to provide clean and healthy potatoes to the user. Defective tubers do not do the industry any good when it comes to consumption of potatoes.

CHAPTER 4

Potato Variety/Cultivars

Potatoes may be classified as follows:

- Smooth Skin
  - White Flesh
    - Round
    - Long or Oblong
  - Yellow Flesh
    - Round
    - Long or Oblong
- Russet Skin
  - White Flesh
    - Round
    - Long or Oblong
  - Yellow flesh
    - Round
    - Long or Oblong

Potatoes for processing are primarily the smooth skin white flesh types except for the dried products wherein the white fleshed russet potato is used extensively. The potatoes are mostly round other than for the French Fry industry wherein long varieties are desirable to meet grade requirements. In Europe and other parts of the World the Yellow fleshed types are desired.

Bob O'Keefe from Nebraska states a chip variety must have two basic characteristics. First, it must be a variety that can be grown profitably by a grower when subjected to the environment and cultural conditions of a given area. This implies it must have: (1) high yielding ability of marketable size and grade quality, (2) disease resistance, (3) maturity to fit the growing season of a given area, and (4) tubers that store well without rotting or sprouting and maintain their chip quality. Secondly, a chipping variety must meet the specifications of the chipper. This implies the variety must produce potatoes that: (1) chip to acceptable color or are low in reducing sugars, (2) produce acceptable chip volume or have high dry matter content (specific gravity) and low percent of peeling waste, (3) are free from blemishes, which requires free from defects, mechanical injury and physiological disorders, and (4) have good texture and flavor of finished products. These same requirements are those needed for most potato cultivars for processing.

Today there are well over 260 different cultivars in use in the potato industry. However, only a few of them are of interest and concern to the chipper. Some of these are shown in the following table.

FIGURE 4.1 — Some Potato Varieties/Cultivars for Chip Usage

Variety/ Cultivar	Specific Gravity	Shape	Count/ 8 lbs.	Storage in °F	Recondition Days at 70°F
Norchip	1.080	Round	24	50-55	10
Atlantic	1.085	Oval-Round	26	50	5
Kennebec	1.080	Oblong	22	55	5
Russett Burbank	1.085	Long	20	55	10
Gemchip	1.080	Round	26	50	5
Steuben	1.083	Round	22	55	5
Somerset	1.080	Oblong	24	50	5
Allegheny	1.080	Round	21	55	5
Snowden	1.083	Round	26	45	10
Monona	1.075	Rd-Oblong	28	45	10

Proper variety selection is most important for the production of quality potatoes that are usable by the processing industry. Some examples of the new cultivars are:

NORCHIP—A release in 1968 from North Dakota State University and developed by Robert Johansen. This cultivar has been one of the better chip varieties. It is characterized by being slightly round to oblong in shape, white skin, shallow eyes and white flesh. It is high in solids and low in sugars if properly matured at harvest. It is an exceptionally good yielder and stores well.

FIGURE 4.2 — NORCHIP (Florida Grown)

SNOWDEN—This cultivar was tested as W855. High specific gravity; and stores well even at 45°F.

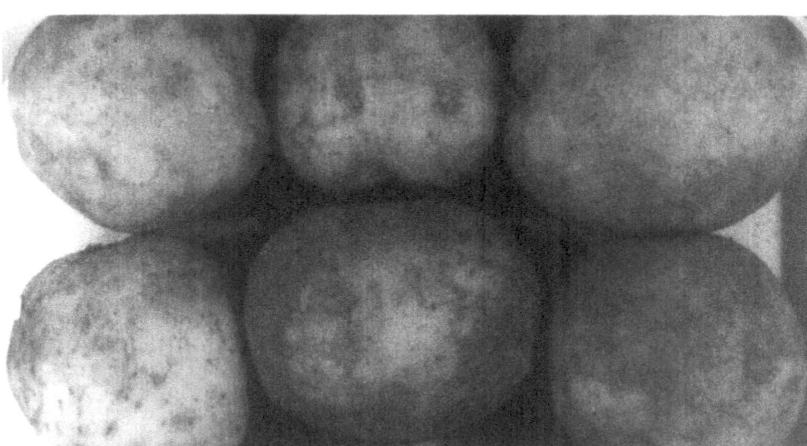

FIGURE 4.3 — SNOWDEN (Florida Grown)

GEMCHIP—A USDA introduction. High yielder and resistance to Verticullium wilt, but susceptible to scab and early blight.

SOMERSET—Released by University of Maine and is a medium to late maturing variety. Average specific gravity, but excellent chip color both before and after storage (45°F). Tubers are oblong and susceptible to hollow heart, Verticullium wilt, scab and sun greening.

FIGURE 4.4 — SOMERSET (Florida Grown)

ATLANTIC—USDA release in 1976. Round with white skin, medium late maturing, very high specific gravity. Susceptible to Hollow Heart, and heat necrosis. Good chipper from the field.

MONONA—Frito Lay release in 1964. Round and white skin with deep eyes but low in specific gravity. Good chipper out of storage

STEUBEN—1988 release from New York Experiment Station. Round, white, medium specific gravity, and late maturing tuber. Large tubers.

ALLEGANY—1988 release from New York Experiment Station. Round white, medium specific gravity, chips well from storage.

FIGURE 4.5 — ATLANTIC (Florida Grown)

Plant breeders from several universities, the USDA and many private firms are developing new cultivars and varieties for use in the potato industry. The Snack Food Association is sponsoring several projects to help the industry help themselves in the production and release of new cultivars for chip use. Many new numbered cultivars are in the testing stage and they include some new "cold chippers", that is, potatoes that can be stored at or near 40°F (4.4°C). The advantage of the colder storage is that diseases are less prevalent, there is less loss of moisture or shrinkage in weight during storage, little or no sprouting and no need for sprout inhibitors. All in all better potatoes for the market by storing at colder temperatures. The future looks bright.

CHAPTER 5

Receiving and Grading for Quality

The processing of potatoes can be divided into the 4 major areas, that is, Canning, Chipping, Drying and Freezing. Each of these preservation methods requires certain preliminary unit operation steps in preparing the potatoes for processing.

Receiving and Grading

When potatoes are delivered to the food plant the receiving clerk should examine the truck or rail car for condition of the vehicle and the overall condition of the load. He should note if the vehicle is clean and sound and the load apears to be in good condition.

FIGURE 5.1 — Inspecting a Pallet Box Of Tubers

He should then sample the load for quality evaluation by selecting at random three or more 10 to 25 pound samples. Care must be taken to insure the use of clean, disease free potatoes. The tubers should be selected at random and be representative of the whole lot in question. The U. S. Department of Agriculture defines "a sample as a representative part or a single item from a larger whole or group presented for inspection or shown as evidence of quality". They recommend that the sample snatcher select 3 representative samples at random from the load by scooping the tubers up rather than hand selecting them. The Grader or Receiving Clerk should ascertain if the potatoes are acceptable or not by determining the pulp temperature, the condition of the load as to any freezing injury and any off-odors, cleanliness of the potatoes including freedom from soil, sprouts, and the soundness, appearance and general firmness of the tubers. (Figure 5.1).

The samples should then be evaluated for specific gravity (see next Chapter), graded for size and absence of external and internal defects. A subsample should be fried in the plant on the conventional fry line or in the laboratory for chip color. If they do not fry satisfactorily, a reducing sugar evaluation should be made. If the tubers are destined for storage at the chip plant a sucrose evaluation must be made to determine if they are acceptable for storage. The reducing sugar level should be below 0.15% preferably at 0% and the sucrose value should be less than 1.50%. If all is satisfactory the load can be weighed in as an acceptable load either for immediate use or for storage. The Delivery Data Form should be completed (see Figure 5.5 at end of chapter).

Sample and Sampling

For grade and quality evaluation, follow the practice as shown in Figure 5.2.

Pulp Temperature and Other Characteristics

Pulp temperature is one of the first characteristics to be determined to give some idea of how the load has been handled. The pulp temperature should be near 68°F (20°C). If the pulp

FIGURE 5.2 — Quality and Grade Evaluation of Potatoes for Chip Use

Load of Potatoes — Select three (3) 25-lb. representative samples at random from each load of potatoes and blend.

A. $\frac{1}{2}$ of Sample

1. Determine Pulp Temperature
2. Wash & Determine % External defects and indicate types
3. Weigh out three 8-lb. samples and determine specific gravity of each lot and average the values
4. Count the number of tubers in each 8-lb. lot and average
5. Size grade a 25-lb sample into: <u>Large</u> (> 3 inches in diameter), <u>Medium</u> (2–3 inches in diameter), <u>B size</u> (< $1\frac{3}{4}$ inches in diameter) and calculate % of each
6. Peel and determine % peel loss
7. Cut the Large tubers and determine % Internal Defects and indicate Types
8. Randomly select 25 tubers and plug them using a $\frac{1}{2}$-inch borer and determine both Glucose and Sucrose values using the YSI or similar method.

B. $\frac{1}{2}$ of Sample

Fry for Chip Quality

- → Evaluate for Color
- → Evaluate for Freedom from Defects:
  - a. % Blisters
  - b. % Minor Defects
  - c. % Major Defects
- → Evaluate for Flavor
- → Moisture
- → Texture
- → Oil Content

temperature is below 50°F (10°C), the load may have to be reconditioned (warmed up to lower the reducing sugar content to less than 0.15%) for a few days before use.

The ***appearance*** of the tubers are of significance to the user. Clean potatoes, uniformity of shape, size, depth of eyes, flesh color, and peel are important in terms of processing.

Size is of significance to determine the usage of any given lot of the potatoes.

FIGURE 5.3 — Sizing Rings

Medium large round tubers are preferred by the chipper since their shape facilitates the removal of the peel more efficiently and with minimal peel loss. On the other hand, the French fry industry prefers the long oblong type of potatoes. In all cases, the processor prefers uniformity within the load. Wide variation in size may require size grading at the factory and in some cases rejection of tubers or cutting the large ones for efficiency in processing. Size may be determined by measuring the intermediate axis of the tuber and classifying the tubers by the size distribution and/or weight in ounces per individual tuber. Raw potato sizes are shown in the following table taken from the USDA Standards for Grades of Potatoes for Processing (see Appendix I). Intermediate axis may be determined by passing the tubers through ring sizes designated at 4 inch, $3^1/_4$ inch, $2^1/_4$ inch, $1^1/_2$ inch designated as Extra Large, Large, Medium, and Small. Samples from given load should be segregated into their sizes and the size calculated as follows: Using a 30 pound sample containing 75 tubers with 9 designated as Small, 45 designated as Medium and 21 designated as Large. Based on the following equation:

$$x\ \%\ \text{each size category} = \frac{\text{No. of tubers in category}}{\text{total no. of tubers}} \times 100$$

Therefore the load consisted of 12% Small, 60% Medium and 28% Large.

FIGURE 5.4 — Size Classification of Potatoes for Processing

Size Classification	Round or Intermediate Min.	Round or Intermediate Max.	Long Varieties Min.	Long Varieties Max.
Extra Large	4 in.	—	16 oz.	—
Large	3 in.	4 in.	10 oz.	16 oz.
Medium	$2^1/_4$ in.	$3^1/_4$ in.	4 oz.	10 oz.
Size B	$1^1/_2$ in.	$2^1/_4$ in.	$1^1/_2$ in.	$2^1/_4$ in.
Size C	1 in.	$1^1/_2$ in.	1 in.	$1^1/_2$ in.
Creamers	$^3/_4$ in.	$1^1/_2$ in.	$^3/_4$ in.	$1^1/_2$ in.

If the tubers are uniform in size, another highly acceptable method of determining the size is the actual count of tubers in the sample, that is, the number of tubers in an 8 pound sample (the weight used for the specific gravity.)

Size is of great significance to the user and as shown in Figure 5.3, size affects surface area, peel loss, no of chips per pound, etc. Surface area is of particular importance when one is frying as in the manufacture of potato chips to control the actual oil content in the finished product since the oil is adsorbed on the surface. It is, also, of great importance to bag fill.

FIGURE 5.5 — Relationship of Tuber Size to Various Physical Constants for Tubers of Different Diameters

Physical Constants	Sphere Diameter of Potatoes in Inches					
	$1^1/_2$	2	$2^1/_2$	3	$3^1/_2$	4
Surface area in sq. inches	7.06	12.6	20.4	28.1	39.2	50.0
Volume in cubic inches	1.76	4.19	9.16	14.13	23.82	33.51
Ratio of area to volume	4.00	3.00	2.50	2.00	1.75	1.50
Approximate no. of tubers/lb.	11	8.0	5.03	2.60	1.80	1.00
Approximate no. of slices/lb.	264	256	201	125	101	64
Approximate no. of tubers/8 lb.	88	64	40	21	14	8
$^1/_{16}$" peel removal (% vol. loss)	21	17.0	14.50	12.0	10.5	9.0

Shape varies widely among the different cultivars, but not generally within a given cultivar. The shape may vary from round, oblong, near flat, elliptical, or long. The old standard cultivar, Kathadin, was typically round and smooth, while the Russet Burbank is classified as long and the Kennebec as elliptical. The round type is preferred because of less loss in peeling and slicing.

Eye depth is an inherited trait and quite uniform within a cultivar. The eye should be shallow for less loss in peeling, particularly when abrasive peeling. Superior cultivar probably has as deep an eye as any cultivar while Atlantic, Somerest, and Snowden are shallow eyed cultivars.

Peel on a tuber develops after suberization and may become quite thick, that is, 8 to 10 cells deep depending on the storage humidity. Peel, also, depends on the potato cultivar, that is, whether or not it is a smooth skinned cultivar or a russet cultivar.

FIGURE 5.6 — Specifications of Potatoes for the Chip Market

CULTIVAR	Norchip, Atlantic, Snowden, Gemchip and Mainechip or equivalent
MATURITY	No feathering - Firm Skin Set
SUGARS: REDUCING SUCROSE	< 0.15%...but, prefer 0.00 < 1.5%...but, prefer 0.00
SIZE	Minimum of 2" in diameter Maximum of 3½"
UNIFORMITY OF SIZE	Maximum variation of ¾"
SHAPE	Round
EYE DEPTH	Shallow
PEEL	Light in color < ⅛" thick
DIRT	Clean and Free of
EXTERNAL DEFECTS	Maximum of 4% bruises, preferably 0% No Soft Rot No Greening No Sprouts No Wire worm No Insect Damage
INTERNAL DEFECTS	No Hollow Heart No Discoloration No Rot No Internal Sprouts
FLESH COLOR	White to light Yellow OR Gold/Yellow
SPECIFIC GRAVITY	> 1.080
TOTAL SOLIDS	> 20.2%
CHIP COLOR	SFA 3 or Lower AGTRON (E30) (90/90) > 45
FLAVOR	Typical and No off-flavors

The total amount of peel loss may be in excess of 20% of the weight of tuber. Peel loss is a function, in part, of tuber size as shown in the tuber size table.

Without doubt, specific gravity is one of the most significant characteristics about the raw potato and it will be discussed in the next chapter because of its importance.

The raw potato is the most significant material in the manufacture of potato chips. The attached specification singles out details of the above and other significant characteristics and attributes when procuring potatoes for chip manufacture.

FIGURE 5.7 — Potato Receiving Inspection Report

Date received: _______ Producer: _______ Variety: ___________
Area: _____________ Load Weight: __________
Carrier: __________ Carrier ID#: __________ Condition of Carrier:
Exterior: Clean: Y/N Repair: Y/N
Interior: Foreign Material: Y/N, Wet: Y/N, Off Odor: Y/N, Rodents: Y/N
Evidence of Chemicals: Y/N, Insects: Y/N, Other: Y/N
Explain ____

Tempt:	Specific Gravity/Solids	No./8#	Size of Tubers $\% < 1\ ^7/_8$	$\% > 3\ ^1/_4$
1.______	1.__________________	________	_______	______
2.______	1.__________________	________	_______	______
3.______	1.__________________	________	_______	______

Tuber Shape: Long: _____ Round: _____ Oblong: _____ Other: _____
% Damaged Tubers: _________ Type: ___________
% External Defects: _________ Type: ___________
% Tubers Greening __________ % Other External Defects _____________
% Internal Defective Tubers _______________ Explain: ________________
% Soil in Load: ______________ Explain: _____________________________
% Reducing Sugar: Sample (1) _____ (2) _____ (3) _____ Average: _______
% Sucrose: Sample (1) _____ (2) _____ (3) _____ Average: _______
Fry Sample: Slice Thickness ______ # 1" _____ Flat: Y/N - Wavy: Y/N
Wash: Y/N, Hold in Minutes: _______ Temperature of Hold Water: _______
Fry Temperature Start: _____ Delta T: _____ End: _____ Seconds: ______
% Moisture: _______ Agtron: _______ SFA Fry Test Color (1-3): ______
# Minor Defects (<1/4" Discolored): _____% of Total Slices: _______
# Minor Defects (<1/2" Discolored): _____% of Total Slices: _______
# Minor Defects (<1/4" Blemished): _____% of Total Slices: _______
# Minor Defects (<1/2" Blemished): _____% of Total Slices: _______
Absence of Defect Score (Maximum of 20 points): ________
Color Score (Maximum of 30 points): ________
Texture Score (Maximum of 20 points): ________
Flavor Score (Maximum of 30 points): ________
TOTAL SCORE:

Accept Load: ______ Reject Load: ________ Disposition: _____________
Inspector: __________________ Date: _________
Comments: ____

CHAPTER 6

Specific Gravity of Potatoes

Specific Gravity is one of the most widely accepted measurements of internal potato quality. Specific gravity is an estimate of the total solids or dry matter content of the tuber. The potato tuber is composed of water (63 to 87%) and (13 to 37 %) dry matter or total solids. The total solids are made up of carbohydrates (13 to 30.5%), protein (0.7 to 4.6%), and mineral or ash (0.50 to 2.00%). The carbohydrates are primarily starch (75-95% amylopectin and 21-25% amylose). Further, the specific gravity of a tuber is an indirect measure of the water content of the tuber.

Specific gravity has a great influence on the texture of the raw potato and products made from potatoes. Specific gravity is associated with dryness and mealiness of baked, boiled, and fried potatoes. For this reason, specific gravity is considered an excellent indicator of final product quality.

To the potato chipper, dehydrator, and French fryer, specific gravity has a direct relationship to the processing efficiency, time and temperature of frying or drying, and yield and recovery of finished products. Higher yields, less cooking oil adsorption, and crisp products are associated with high specific gravity potatoes. It stands to reason, since the frying operation is essentially a drying operation, that the more moisture or water in the potato or the less total solids to start with, the less solids in the end product and the more energy needed to remove the moisture with a greater oil adsorption in the final product. Specific gravity should be in excess of 1.080 for these types of products.

Specific gravity is just as important to the salad manufacturer, canner, or the grill cook as low specific gravity potatoes generally will not slough, become soft, or loose their texture nearly as quick as high specific gravity potatoes. Here, specific gravity should be less than 1.070 for these types of products, preferably 1.060 or lower for best qualities

FIGURE 6.1 — Scale and basket of potatoes.

FACTORS INFLUENCING SPECIFIC GRAVITY

Many research workers and writers have covered the subject of factors influencing specific gravity. There is some confusion on what the authorities have to say probably due to the different type of experiments the research workers have conducted or the different set of conditions in which they have conducted their experiments including their respective areas of production. The following comments are taken from some of these authorities along with my own experiences. I have attached to this article many references. The details of many studies may be found in this list of references.

Some of the ***uncontrollable*** factors which may cause stress during growth and ultimately have an effect on the specific gravity of potatoes include:

Climate and Growing Season

Date of planting is dictated by the date of the last frost and the date of harvest which may be dictated by the first frost. Both are critical factors in some areas of potato production. The frost free days are defined as the maximum length of the growing season. If the season is too short, the harvested tubers may not have time to reach maximum growth and thus the specific gravity and yield may be lower than normal.

Day Length and Light Intensity

Day length and light intensity have been postulated as a significant factor for development of the tuber and the ultimate level of the specific gravity, particularly for certain varieties grown in the northern tier of states. We know that sunshine is required for photosynthesis to take place in the leaves of the potato plant. If sunshine is limited due to short days, less photosynthetic activity and therefore, less carbohydrates are being manufactured. Associated with sunshine and day length, the potato plant needs cool nights for maximum reduction in the respiration of the plant as respiration consumes a lot of the daily build up of carbohydrates and, thus, the carbohydrates never get translocated to the tuber and the total solids or specific gravity does not increase. It is a well known fact that plants do not mature with long day lengths as well as they do when grown with short day lengths. In many cases potatoes grown under day length conditions are dug premature and they will not store for any period of time. They may accumulate good specific gravity if all other conditions are conducive for growth, but they seldom mature and thus are not good stored potatoes.

Air Temperature

This has been reported as a major factor as it affects the plant growth and tuberization. All plants have specific growth temperatures, that is minimum and maximums. Minimum temperature for a potato plant has been reported to be 38°F and the maximum for manufacture of carbohydrates is variable due to other associated factors, that is moisture and sunlight. Some state that if the temperature exceeds 70°F, the accumulation of carbohydrates in the tuber is retarded, thus causing lower specific gravity. As indicated above, high night temperatures during the growing season causes high respiration rates and thus there is little accumulation of carbohydrates in the tuber. Some authors indicate that temperature in excess of 95°F are detrimental unless irrigation programs are in place and in use.

Wind and Humidity

These two factors have been reported as factors affecting specific gravity and they do have great effect on stress due to high evaporation of moisture from the plant and soil and, thus, do not allow for accumulation of solids or specific gravity buildup in the tuber.

Drought

Water is essential for potato plant growth. One researcher has stated that potatoes require some 400 pounds of water to synthesize one pound of dry matter. He calculated that for a growing season of 120 days and for a yield of 500 bushels to the acre, 18 acre inches of rainfall are required. He further stated, that water levels above or below this amount would have a direct bearing on lower specific gravities of tubers. This same author states that moisture applied at blossom time presumably sets the stage for the production of a large number of tubers. Moisture applied following the peak of physiological maturity might well lower the specific gravity of developing tubers and cause secondary growth of tubers. The critical period of moisture supply for high specific gravity is during the period when tubers are rapidly enlarging and approaching maturity. Excess moisture before or after this period lowers specific gravity.

High Soil Temperature

Generally this is not a factor if the grower has no exposed potato hills, but it is important to make certain that the top of the row does not exceed three inches above ground level for proper root growth and adequate moisture retention. Soil temperature has a great effect on storability of tubers for certain markets. Most workers indicate that tubers should be harvested when the soil temperature is, at least, above 45°F. Some workers indicate that it should never be below 50°F for bruise prevention and the prevention of reducing sugar accumulation due to respiration. One further comment concerns tubers growing at relatively high uniform temperatures and then harvested with a drop in temperature, studies indicate that the chips may fry dark. It should be fairly obvious that the reducing sugars have not been converted to sucrose and sucrose to starch fast enough and the excess amount causes chip darkening.

No doubt there are other uncontrollable factors, but these are some of the ones that man has little or no control over other than using timely management practices. The ***controllable*** factors may have more of an impact on specific gravity. One author has stated that 90% of the specific gravity problems can be attributed to one or more of the following controllable factors:

Variety/Cultivar

Exhaustive studies in the past and much emphasis today would indicate that this is the most single factor affecting specific gravity. Not all varieties or cultivars are adapted to all areas of production and, therefore, it is not possible to state that one should only grow a specific cultivar or variety. Some cultivars or varieties are known as late maturing cultivars and they may never mature in certain potato growing areas. Further, not all cultivars or varieties react the same every year. The data in Figure 6.2 illustrate this point. It is possible to state that the use of specific cultivars in recent years have shown some compliance

FIGURE 6.2 — What Does Specific Gravity Really Mean?

Assume we have two plants with equal equipment and capacities, but each uses potatoes having different specific gravities. **The rule is that each .005 increase in specific gravity equals 1 lb. increase in yield of chips.**

Plant A uses:	Plant B uses:
Sp. Gr. 1.070	Sp. Gr. 1.095

A difference of .025, or 5 more lbs. of chips per 100 lbs. of raw potatoes. Therefore, if each plant runs 10,000 pounds of potatoes:

	Daily production =	
3,000 lbs.		3,500 lbs.
	Weekly production =	
15,000 lbs.		17,500 lbs.
	Or in 50 Weeks	
750,000 lbs.		875,000 lbs.

Assuming chips worth $3 per lb.; therefore, production valued at:

$2,250,000.00 verses $2,625,000.00

In addition, less energy to fry equivalent pounds of potatoes, and the finished chips will have significant less oil content.

---

with desired specific gravities. Norchip from the Red River Valley area has generally had very high specific gravities over the years. Atlantic from Florida and other areas has had exceptionally high specific gravities averaging in the 1.085 range year after year. Many other cultivars look very promising for the proper development of specific gravities in several areas.

Seed Quality and Seed Piece Size

As with any crop, what you plant is what you get. Poor quality seed and small seed pieces are not conducive to high specific gravities. Some authors indicate that a minimum seed piece should be one and three fourths ounces and preferably up to 3 ounce size.

Planting Density

The number of plants per acre have a direct relationship with cultivar or variety, soil fertility, water usage, etc. There are no easy answers to plant population. Generally, local experience with specific cultivars, etc. dictates plant population.

Mineral Nutrition and Fertilizer Practices

This is a topic of great interest and much discussed in the literature. There are some agreements and the first is that plants must have nutrients for growth and secondly that excess nitrogen and potash can enhance vine growth and retard maturation of the plant and consequently lower yields and specific gravity. The well managed potato grower relies on soil and leaf analysis for maximizing his efforts with his crop. Further, he must tie mineral nutrition and fertilization practices to day length and light intensity as they relate to fertilization programs for any given area of production.

Soil Type

There are wide differences of opinion varying from some success with muck soil types to excellent results on upland soil. Some prefer sandy and heavy clay soil over well textured soils. Most authors agree that potatoes can be gown on most soil types if the soils are not compact and if they have good texture with optimum nutrients and water supply available for the growing plant. However, the potato plant has high oxygen requirements in the root zone when compared to many other agronomic crops and well textured soil is vital for proper growth.

Soil Moisture

This subject was covered under drought above, but the following statements may be helpful and that is, optimum yields are obtained when the soil water is kept above 65% available moisture throughout the growing season. Periods of water stress, coupled with high air temperatures, are responsible for most of the quality problems. In many potato growing areas this implies that one cannot rely on rainfall and must supplement moisture through the use of irrigation.

Pests and Pest Management (Insects, Weeds and Diseases)

This is without question a major area of concern, but modern Integrated Pest Management (IPM) practices and programs have shown us that pests can be controlled. The subject may demand more attention in the next century and the years beyond as there is no question in my mind that man must find genetic differences in potatoes and better use of pest management programs to reduce or eliminate the use of chemicals in production and processing programs. We have become too dependent upon chemicals. We must retract and look at what we are doing to our soils, rivers, and our food supply. I am not an advocate of no chemicals, but I am an advocate of proper use of safe chemicals. The ALAR and EBDC situations reemphasize this fact.

Tuber Maturity

This is a topic of wide differences of opinions and a topic that is not always understood. One must realize that in a growing plant if all conditions are right, (adequate sunlight, a green healthy plant, proper temperature, a supply of water and carbon dioxide) that that plant will continue to manufacture simple carbohydrates and that those carbohydrates will be translocated to the storage organs, that is, the tubers. Thus, a potato plant may never die or at least it will not mature its tubers for some time. To mature potatoes naturally, one must use cultivars or varieties that will grow and mature under a short number of growing days for any given production area or one must rely on man-made methods of killing the potato plant. Potatoes will not mature as long as that plant is growing, that is, carbohydrates are being synthesized and stored as simple carbohydrates until nature can convert them to starch. This process of killing vines was practiced in the early days by allowing disease to take over, or insects to consume the leaves and plants. Modern man has used vine killer chemicals or vine mowing or vine beating. The object has been to stop the manufacture of carbohydrates and allow those that are manufactured to become translocated and stored as starch. In some varieties and in some areas, maturity of plants is a natural phenomenon, in other areas it is quite synthetic or manmade. Regardless, tubers must be mature if they are to be stored for many markets, including the chip industry. Immature tubers can be directly used by some manufacturers because the plant is respiring and the simple reducing sugars are being burned off rapidly and do not cause a discoloration problem in the manufacturing process.

Man knows today how to predict maturity by measuring sucrose, the intermediate carbohydrate in the manufacturing process. As long as the sucrose content is below 1.5% percent of the dry matter content, tubers can be harvested and stored and generally they will make satisfactory potato chips and dried products if the glucose content is less than 0.05% of the dry matter content of the tubers. One must remember that tubers at harvest and during their storage life are living tissues and chemical processes may continuously be going on inside the tuber until the tuber, a living tissue, is killed. This is done by inactivating the natural occurring enzymes during the processing of potatoes into manufactured shelf stable products. Therefore, the timing of the harvest in terms of maturity is critical depending on the ultimate use of the crop. Growers must know their markets for handling potatoes wisely.

DISCUSSION OF SPECIFIC GRAVITY

Specific gravity measurement is a relatively simple and efficient method for estimating the total solids content of potatoes. Although high correlation's have been found between specific gravity, dry matter and starch content, variation of the estimated total solids as high as plus or minus 4.5% from the actual values have been reported. Numerous researchers have identified sources of the additional variation in the relationship between specific gravity and dry matter content for potato tubers. One researcher has stated that deviations from the specific gravity and dry matter relationships was attributable to proportional differences in the air volume in the tuber, that is, in the intercellular airspace's. Another worker suggested that this variation within a sample of a given cultivar was an inherent trait and cultivars with low variation tended to remain known from year to year. Several workers have calculated regression equations between specific gravities and dry matter content.

Specific gravity of potatoes has been shown to directly influence the yield and quality of potato products. Further, it has proved to be a reliable indicator of oil adsorption of finished chips.

FIGURE 6.3 — Conversion Table for Specific Gravity of Potatoes to Water, Dry Matter, Chip Yield and Percent Oil in Two Styles of Chips

Specific Gravity	% Water	% Dry Matter	% Chip Yield	% Oil in Chips	
				Flat	Wavy
1.050	86.10	13.88	27.00	44.5	43.0
1.055	85.06	14.94	27.75	44.0	42.5
1.060	84.01	15.99	28.50	43.5	41.8
1.065	82.95	17.05	29.25	42.5	41.0
1.070	81.90	18.10	30.00	41.6	40.0
1.075	80.84	19.16	30.75	40.8	39.0
1.080	79.79	20.21	31.50	39.8	38.0
1.085	78.73	21.27	32.31	38.5	37.3
1.090	77.67	22.33	33.10	37.5	36.6
1.095	76.62	23.38	33.90	36.5	35.8
1.100	75.56	24.44	34.70	35.8	35.0
1.105	74.51	25.49	35.45	35.1	34.0
1.110	73.45	26.55	36.25	34.3	33.1

MEASUREMENT OF SPECIFIC GRAVITY

Several methods are available for the measurement of specific gravity, although not all have the same degree of precision. Several significant points must be remembered when measuring the specific gravity of tubers. The first is that the temperature of the tuber or pulp temperature and the temperature of the water or solution used in determining the specific gravity of the tubers must be the same. If these temperatures are not the same, correction factors given in Figure 6.4 should be used for adjusting the specific gravity.

Secondly, tubers must be clean, free from disease, and hollow heart and dry for measuring their specific gravity. Further, bruised tubers should not be used.

Thirdly, Tubers from any given hill, load, or lot will vary in their specific gravity content by some 0.006 above or below the average specific gravity for the sample. Sometimes, these values may exceed this level. Proper sampling and the measurement of 3 or more samples per lot is essential for a predictable specific gravity of any given lot. An average specific gravity of 1.080 simply means that one-half of the tubers in that sample will have a low specific gravity of 1.074 and the other half will be as high as 1.086. If the sample was not carefully taken from the lot and be representative of the whole, the readings could be quite misleading and unreliable.

Air and Water or Under Water Method

This method of determining specific gravity is based upon Archimedes principle, that is,

$$\text{Specific Gravity} = \frac{\text{Weight in air}}{(\text{weight in air})\ (\text{weight in water})}$$

This method is accurate, simple, and convenient. It has the advantage in that one can determine the specific gravity of individual tubers or the specific gravity of a large lot depending on one's scale capacity. For accurate measurement, the weight in water is most significant and one must use an accurate scale for this measurement.

FIGURE 6.4 — Conversion Table for Specific Gravity of Potatoes to Water, Dry Matter and Starch in Percent

(Calculated from Von Scheele equations: % starch = 17.564 + 199.07 (Sp. Gr. – 1.0988; % Dry Matter = 24.182 + 211.04 (Sp. Gr.– 1.0988; % Water by difference or 100 – % Dry Matter; % Yield calculated from assuming 15% loss in peeling, washing, sorting, and trimming) and % oil calculated for Flat Chips sliced at 0.063" and fried with Inlet Oil at 375°F and Outlet Oil at 345°F and Wavy chips sliced at 0.075" and fried at 375°F Inlet Oil and finished at 345°F Outlet Oil.

Specific Gravity	% Water	% Dry Matter	% Starch	% Chip Yield	% Oil in Chips Flat	% Oil in Chips Wavy
1.050	86.12	13.88	7.85	27.00	44.5	43.0
1.051	85.91	14.09	8.05	27.15	44.4	42.9
1.052	85.69	14.31	8.25	27.30	44.3	42.8
1.053	85.48	14.52	8.45	27.45	44.2	42.7
1.054	85.27	14.73	8.65	27.60	44.1	42.6
1.055	85.06	14.94	8.85	27.75	44.0	42.5
1.056	85.15	15.15	9.04	27.90	43.9	42.4
1.057	84.62	15.38	9.24	28.05	43.8	42.2
1.058	84.43	15.57	9.44	28.20	43.7	42.1
1.059	84.22	15.78	9.64	28.35	43.6	42.0
1.060	84.01	15.99	9.84	28.50	43.5	41.8
1.061	83.79	16.21	10.04	28.65	43.3	41.6
1.062	83.58	16.42	10.24	28.80	43.1	41.5
1.063	83.37	16.63	10.44	28.95	42.9	41.3
1.064	83.16	16.84	10.64	29.10	42.7	41.2
1.065	82.95	17.05	10.84	29.25	42.5	41.0
1.066	82.74	17.26	11.04	29.40	42.3	40.8
1.067	82.53	17.47	11.23	29.55	42.1	40.6
1.068	82.32	17.68	11.43	29.70	42.0	40.4
1.069	82.11	17.89	11.63	29.85	41.8	40.2
1.070	81.90	18.10	11.83	30.00	41.6	40.0
1.071	81.68	18.32	12.03	30.15	41.5	39.8
1.072	81.47	18.53	12.23	30.30	41.3	39.6
1.073	81.26	18.74	12.43	30.45	41.2	39.4
1.074	81.05	18.95	12.63	30.60	41.0	39.2
1.075	80.84	19.16	12.83	30.75	40.8	39.0

FIGURE 6.4 — Continued

Specific Gravity	% Water	% Dry Matter	% Starch	% Chip Yield	% Oil in Chips Flat	Wavy
1.076	80.63	19.37	13.03	30.90	40.6	38.8
1.077	80.42	19.58	13.22	31.05	40.4	38.6
1.078	80.21	19.79	13.42	31.20	40.2	38.4
1.079	80.00	20.00	13.62	31.35	40.0	38.2
1.080	79.79	20.21	13.82	31.50	39.8	38.0
1.081	79.57	20.43	14.02	31.66	39.5	37.8
1.082	79.36	20.64	14.22	31.82	39.3	37.6
1.083	79.15	20.85	14.42	32.00	39.0	37.5
1.084	78.94	21.06	14.62	32.16	38.8	37.4
1.085	78.73	21.27	14.82	32.31	38.5	37.3
1.086	78.52	21.48	15.02	32.47	38.3	37.2
1.087	78.31	21.69	15.22	32.63	38.1	37.0
1.088	78.10	21.90	15.41	32.79	37.9	36.9
1.089	77.89	22.11	15.61	32.94	37.7	36.8
1.090	77.67	22.33	15.81	33.10	37.5	36.6
1.091	77.46	22.54	16.01	33.26	37.3	36.4
1.092	77.25	22.75	16.20	33.42	37.1	36.2
1.093	77.04	22.96	16.41	33.58	36.9	36.0
1.094	76.83	23.17	16.61	33.74	36.7	35.9
1.095	76.83	23.38	16.81	33.90	36.5	35.8
1.096	76.41	23.59	17.01	33.06	36.4	35.6
1.097	76.20	23.80	17.21	33.22	36.2	35.5
1.098	75.99	24.01	17.41	33.38	36.1	35.3
1.099	75.78	24.22	17.60	33.54	35.9	35.2
1.100	75.56	24.44	17.80	34.70	35.8	35.0
1.101	75.35	24.65	18.00	34.85	35.7	34.9
1.102	75.14	24.86	18.20	35.00	35.5	34.7
1.103	74.93	25.07	18.40	35.15	35.4	34.5
1.104	74.72	25.28	18.60	35.30	35.2	34.3
1.105	74.51	25.49	18.80	35.45	35.1	34.0
1.106	74.30	25.70	19.00	35.61	34.9	33.8
1.107	74.09	25.91	19.20	35.77	34.7	33.6
1.108	73.88	26.12	19.40	35.93	34.6	33.4
1.109	73.66	26.34	19.60	36.09	34.5	33.3
1.110	73.45	26.55	19.79	36.25	34.3	33.1

The Brine Method

This method is based on the fact that potatoes will float or sink in a solution, depending on the density of the solution. Usually a series of solutions are made up (Figure 6.5) and the tubers are moved from one density to another to determine which solution they just float in. This method is relatively slow, even after the solutions are made up. However, it is accurate if the

FIGURE 6.5 — Specific Gravity, Percent Salt, Degree Salometer and Ounces of Salt Per Gallon Relationships

Specific Gravity	Percent NaCl	Salometer Reading	Ounces of Salt
1.040	5.8	21.9	8.5
1.045	6.5	24.4	9.33
1.050	7.1	26.9	10.51
1.055	7.8	29.4	11.70
1.060	8.5	31.9	12.88
1.065	9.1	34.4	14.06
1.070	9.8	36.9	15.24
1.075	10.4	39.3	16.42
1.080	11.1	41.8	17.60
1.085	11.8	44.3	18.79
1.090	12.4	46.8	19.97
1.095	13.1	49.3	21.15
1.100	13.7	51.8	22.33
1.105	14.4	54.3	23.51
1.110	15.1	56.8	24.70
1.115	15.7	59.3	25.88
1.120	16.4	61.7	27.06
1.125	17.0	64.2	28.24
1.130	17.7	66.7	29.42
1.135	18.3	69.2	30.61
1.140	19.0	71.7	31.79

operator is careful in determining which density the tuber floats in a 30-second time frame. It is generally used for measuring the specific gravity of individual tubers and thus has not found as wide an application as one would hope. Large-scale brine separators have been constructed and they do work, but marketing specific gravity lots has not justified the expenditure of this equipment or the salt for commercial separation applications. Once the tuber is wet, it must be used or disease may become a problem.

Hydrometer Method

This is a patented hydrometer developed and sold by the Snack Food Association. Basically, it is a large hydrometer with a hook on the bulb at the bottom to attach a basket containing 8 pounds of potatoes. The hydrometer can be carefully calibrated by using a steel rod and adjusting the scale inside the tuber of the hydrometer. One must use exactly 8 pounds of potatoes or the readings are not accurate. Further, one must have a container at least 40 inches deep and large enough to allow the basket to float freely without touching or being impeded by the sides. The method does require a scale to weight exactly 8 pounds in addition to the basket weight. For accurate readings, the instrument should be accurately calibrated and the tubers should be clean and free of any disease or Hollow Heart, as previously mentioned. Ideally, the water and pulp tempeature should be the same or corrections must be made as shown in Figure 6.6. The relationship of specific gravity to % water, % starch and % dry matter are shown in Figure 6.7.

In summary, variations in specific gravity of potatoes do occur, but much of the variation in specific gravity is controllable. Depending on the use of potatoes, specific gravity should be a part of the specifications for the various outlets for fresh potatoes. Generally, the potato salad market desires potatoes with specific gravities less than 1.075, the baking trade desires potatoes with specific gravities that exceed 1.090, the french fry market desires potatoes with specific gravities between 1.075 to 1.095, the flake or dehydration market and the chip market desire potatoes with specific gravities from 1.080 to 1.100. Workers at Cornell

FIGURE 6.6 — Correction Factors for Specific Gravity of Potatoes
Corrected to Zero Base of 50°F Tuber Temperature and 50° F Water Temperature

Tuber	Water Temperature (° F)									
Temperature	30°	40°	45°	50°	55°	60°	65°	70°	75°	80°
38°	–0.0021	–0.0020	–0.0018	–0.0018	–0.0020	–0.0023	–0.0029	–0.0038	–0.0047	–0.0056
40°	–0.0017	–0.0016	–0.0014	–0.0014	–0.0016	–0.0019	–0.0034	–0.0034	–0.0043	–0.0052
45°	–0.0009	–0.0008	–0.0006	–0.0006	–0.0008	–0.0011	–0.0017	–0.0026	–0.0035	–0.0044
50°	–0.0003	–0.0004	–0.0000	–0.0000	–0.0002	–0.0005	–0.0011	–0.0020	–0.0029	–0.0038
55°	+0.0001	+0.0002	+0.0002	+0.0002	–0.0001	–0.0001	–0.0007	–0.0016	–0.0025	–0.0034
60°	+0.0004	+0.0005	+0.0007	+0.0007	+0.0005	+0.0002	–0.0004	–0.0013	–0.0022	–0.0031
65°	+0.0005	+0.0006	+0.0008	+0.0008	+0.0006	+0.0003	–0.0003	–0.0012	–0.0021	–0.0030
70°	+0.0006	+0.0007	+0.0009	+0.0009	+0.0007	+0.0004	–0.0002	–0.0011	–0.0020	–0.0029
75°	+0.0007	+0.0008	+0.0010	+0.0010	+0.0008	+0.0005	–0.0001	–0.0010	–0.0019	–0.0028
80°	+0.0008	+0.0009	+0.0011	+0.0011	+0.0009	+0.0006	+0.0000	–0.0009	–0.0018	–0.0027
85°	+0.0009	+0.0010	+0.0012	+0.0012	+0.0010	+0.0007	+0.0001	–0.0008	–0.0017	–0.0026
90°	+0.0010	+0.0011	+0.0013	+0.0013	+0.0011	+0.0008	+0.0002	–0.0007	–0.0016	–0.0025
95°	+0.0011	+0.0012	+0.0014	+0.0014	+0.0012	+0.0009	+0.0003	–0.0006	–0.0015	–0.0024
100°	+0.0012	+0.0013	+0.0015	+0.0015	+0.0013	+0.0010	+0.0004	–0.0005	–0.0014	–0.0023

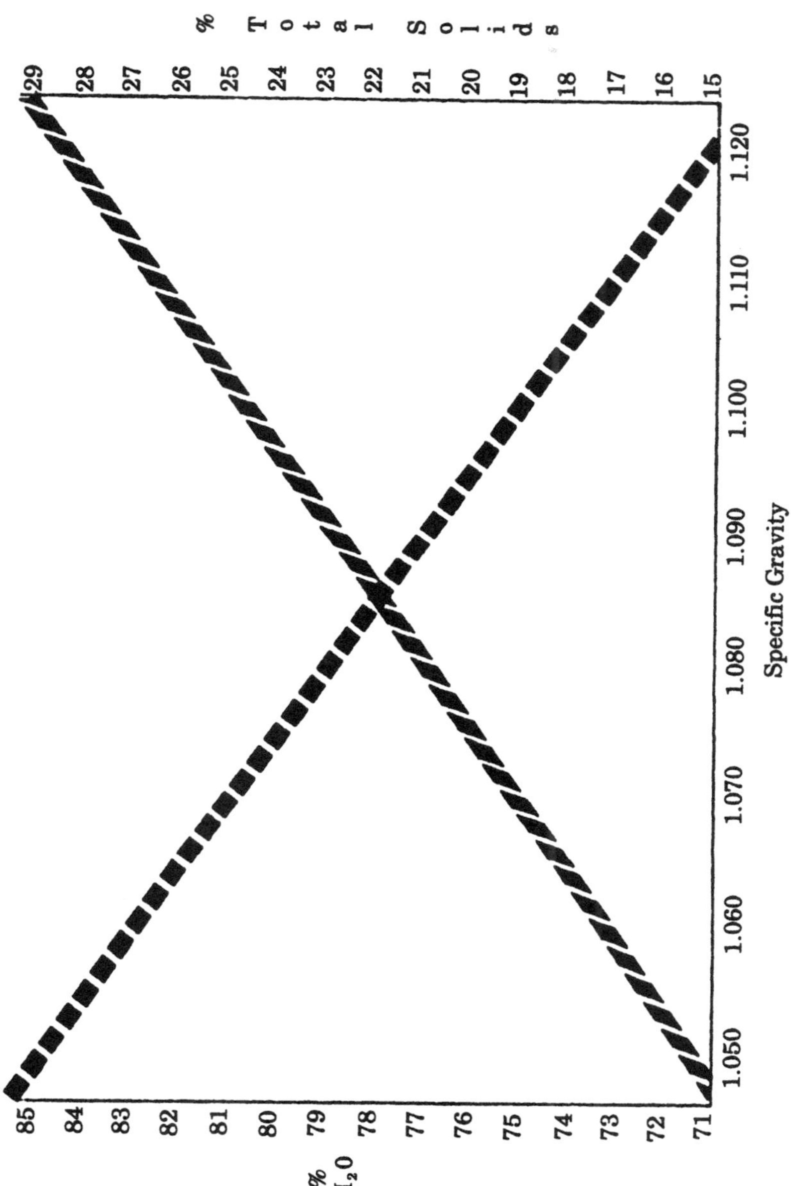

FIGURE 6.7 — Relationship Between % Water, % Total Solids and Specific Gravity

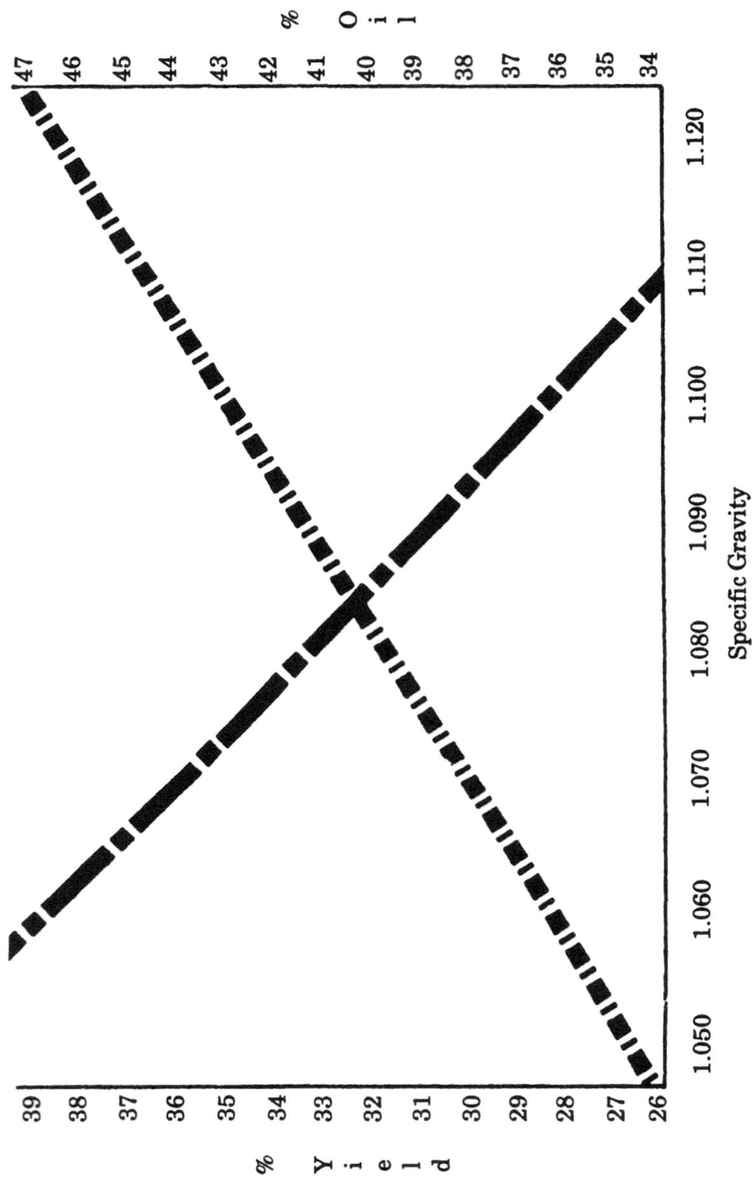

FIGURE 6.8 — Relationship Between % Yield, % Oil and Specific Gravity

FIGURE 6.9 — Relationship Between % Oil and Specific Gravity

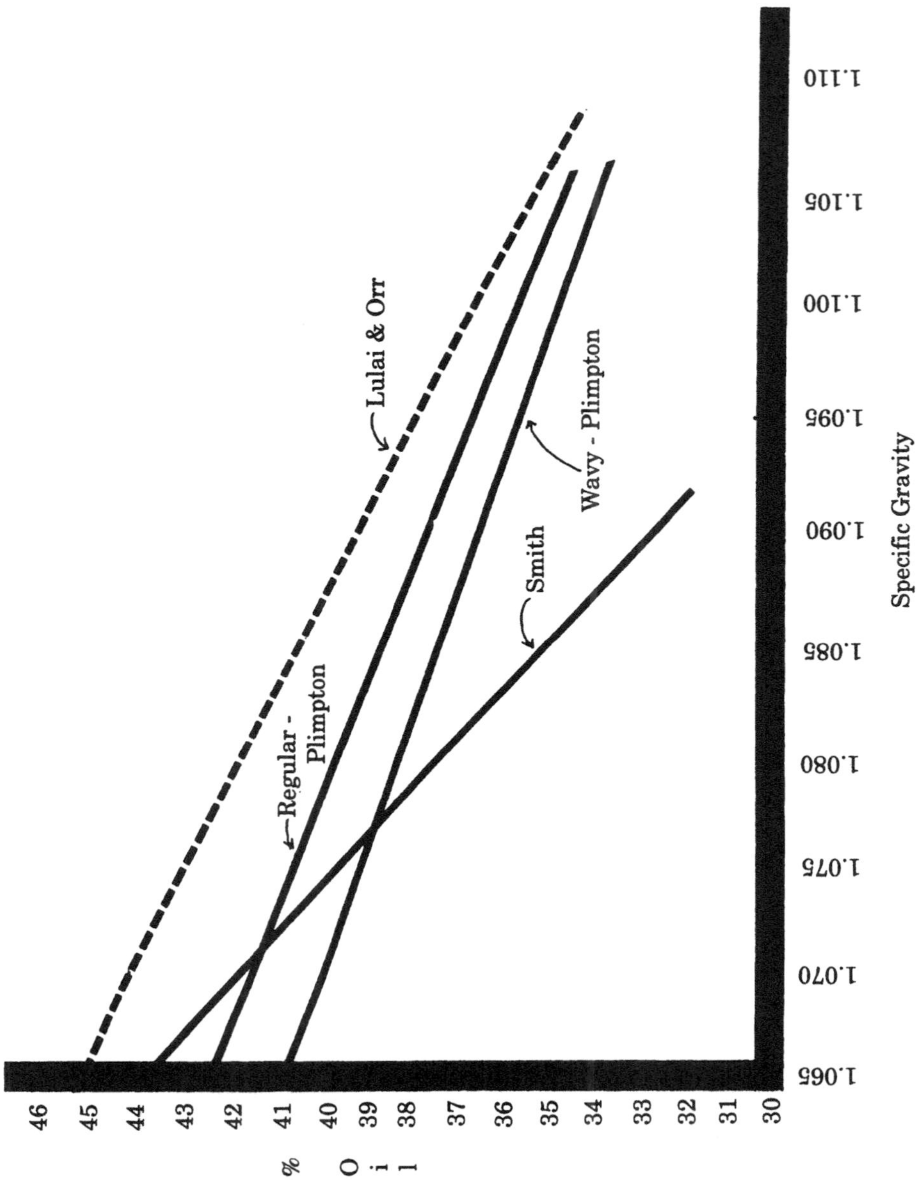

Univesity report that for every 0.005 increase in specific gravity, one can expect an extra pound of chips per 100 pounds of raw potatoes. (Also note the relationship of specific to gravity to total solids as shown in Figure 6.9.) Specific gravity is easy to determine and knowledge of it should be a major selling point in the marketing of fresh potatoes. Specific gravity will remain fairly constant during the normal storage life of the potato if the relative humidity remains high (95%)

FIGURE 6.10 — Testing Specific Gravity of Tubers

FIGURE 6.11 — Prediction Equations, Correlation Coefficients, and 95% Confidence Limits for the Regression of Total Solids on the Specific Gravity of Potato Tubers

Equation (% Total Solids =)	R	95% Confidence Limit	Equation Number
3.33 + 211 (specific gravity – 1)[a]	0.937	± 2.00	(1)
–194.84 + 199.63(specific gravity)[b]	****	± 1.54	(2)
–196.98 + 201.72(specific gravity)[c]	****	± 2.11	(3)
–217.2 + 221.1(specific gravity)[d]	0.912	± 4.47	(4)

[a] From Scheele et al. (1937) using European cultivars.

[b] From Porter et al. (1964b) using five varieties from six different areas in the United States for the 1962 and 1963 growing seasons.

[c] From Fitzpatrick et al. (1969) using 483 tubers from 103 varieties grown in eight areas in the United States during the 1962, 1963 and 1968 growing seasons.

[d] From Schippers (1976) using 1269 tubers from at least 41 cultivars, stored under varying storage conditions grown on Long Island during the 1969, 1970, 1971, and 1972 growing seasons.

FIGURE 6.12 — Relationship of Specific Gravity to Oil Content of Potato Chips

Specific Gravity	% Oil Values Reported By: Smith	Lulai & Orr	Plimpton (OSU) Regular	Wavy
1.065	45.3	45.71	42.5	41.2
1.070	38.0	44.38	41.5	39.8
1.075	36.7	43.05	40.6	38.3
1.080	36.1	41.72	39.6	37.8
1.085	35.9	40.38	38.6	37.2
1.090	32.3	39.06	37.6	36.7
1.095		37.73	36.6	36.2
1.100		37.40	35.6	34.9
1.105		35.07	34.6	33.5

Figure 6.13 — Average Specific Gravity of Five Selected Cultivars Grown on Six Farms in Ohio, 1977–1986

	Cultivars				
Years	Katahdin	Kennebec	Atlantic	Denali	Norchip
1977	1.066		1.072		1.066
1978	1.067		1.085	1.086	1.077
1979	1.060		1.075		1.070
1980	1.062	1.065	1.077	1.078	1.070
1981	1.069	1.079	1.087	1.083	1.076
1982	1.071	1.075		1.088	1.076
1983	1.069	1.075	1.079	1.083	1.072
1984	1.069				1.072
1985	1.072	1.071	1.086		1.070
1986	1.070		1.093		1.070
Average	1.065	1.0730	1.0818	1.0836	1.0725
Range +/–	.006	.007	.0055	.005	.005

CHAPTER 7

Potato Manufacture

Potato Processing

There are several unit operations in the manufacture of potatoes and potato products as shown in the following flow chart. Many of the unit operations are the same for the preparation of potatoes for processing, however the processing methods are quite different. The processing methods will be presented separately after a discussion of the preparation methods. The following processing (preservation) methods will be presented, that is, canning, chipping, drying, and freezing.

Each unit operation in the preparation of potatoes for processing is a separate function in the processing of potatoes. There are specific parameters about each unit operation that the operator must be familiar with for producing high quality and uniform products.

- Potatoes are processed to eliminate waste and prevent spoilage.
- Potatoes are processed to make them more readily available and in convenient form for use.
- Potatoes are processed to increase their value and enhance their worth.
- Potatoes are processed to develop new products, both in types and styles.
- Potatoes are processed to increase their usefulness, extend their shelf life, improve their quality, and preserve their nutrients.

Receiving and Grading

When potatoes are delivered to the chip plant the receiving clerk should examine the truck or rail car for condition of the vehicle and the overall condition of the load. He should ascertain if the potatoes are acceptable or not by determining the pulp temperature, the condition of the load as to any freezing injury and any off-odors, cleanliness of the potatoes including freedom from soil, sprouts, and the soundness, appearance and general firmness of the tubers. He should note if the vehicle is clean and sound. He should then sample the load for quality evaluation by selecting at random three or more 10 to 25 pound samples.

These samples should then be evaluated for specific gravity, graded for size and absence of external and internal defects. For potato chips, a subsample should be fried in the plant on the conventional fry line or in the laboratory for evaluation of the chip color. If they do not fry satisfactorily, a reducing sugar evaluation should be made. If the tubers are destined for storage at the chip plant a sucrose evaluation must be made to determine if they are acceptable for storage. The reducing sugar level should be below 0.15% preferably at 0% and the sucrose value should be less than 1.50%. If all is satisfactory the load can be weighed in as an acceptable load either for immediate use or for storage. The Delivery Data form should be completed (see Chapter 5).

Preparation for Processing

Potatoes for processing are all prepared in much the same manner, that is, after receiving they all are dry sorted, destoned, washed, peeled, and further sorted and trimmed. Each unit operation is a separate function with specific parameters that the operator must be familiar with for producing high quality uniform products. Potatoes move into the processing plant from storage or direct from the delivery truck or rail car or from holding bins as shown in Figure7.1. They may be delivered to the processing plant in crates, by flumes, or by conveyor.

Potatoes may be conveyed or flumed into the manufacturing plant. Fluming is preferred to direct conveying in that the soil can be separated from the tubers before going further in the

FIGURE 7.1 — Holding bins for potatoes.

process line. The flumes for potatoes may contain detergents to help separate the soil from the tubers. Detergents act as wetting agents and tend to help loosen the soil. In the flume, agitation of the tubers can be beneficial. Agitation can be accomplished by bubbling high pressure air in the flume water. If the tubers are cold and if the soil is excessive, hot water can be used to agitate the tubers by injecting steam into the flume water.

Dry Sorting

Dry sorting is the first unit operation wherein the defective potatoes, harmless extraneous matter, trash and containers and other unwanted materials are removed. Most of this is done by inspection of the product as it is being unloaded or before the product starts through the processing line.

Magnetic and electronic metal detectors are being used. The electronic metal detectors use a high frequency electromagnetic field and metals passing through this field disturb it, that is, the metal creates a circuit imbalance and this triggers a reject system, sets off an alarm, or other appropriate signal. The advantage of the system includes its ability to: (a) detect both ferrous and non-ferrous metals; (b) distinguish between product and actual metal debris; and (c) detect metal in varying size ranges. The equipment can be mounted on the line or in the line.

Plastic is very difficult to remove and will create a problem if not picked out of the potatoes. It usually gets in the product when harvesting the potatoes or if plastic is used in handling the potatoes.

Wood can be floated off if present and if not removed will create problems in slicing and frying as in potato chips.

Destoning

Potatoes are destoned to remove any stones that may be in the load. Stones cause serious problems in the slicing operation and must be removed by floating the potatoes in water wherein the stones will go to the bottom and the tubers will float. This may be done with vertical screws or in riffle type washers. Some growing areas are notorious for stones, particularly virgin fields. Some metal can be removed at the time of stone removal. In addition to stones and heavy metal, wood, plastic, aluminum cans, glass etc. must be removed. Usually this is accomplished by visual examination or inspection of the tubers as they are unloaded onto conveyor belts or into flumes prior to the next unit operation. Wood, particularly from virgin soils can be removed by flotation. Metal detectors may, also, be used to remove the metals if present.

Washing

Potatoes may be washed, if necessary, before further processing to remove sand, dirt, mud, and other extraneous matter. Washing is accomplished with water delivered from high pressure sprays over the tubers while they are being conveyed by rollers onto the next operation. If the tubers were flumed into the plant, the soil

FIGURE 7.2 — De-Stoner/Elevator Hopper Feeder Crate Dumper (Courtesy Heat and Control, Inc.)

may be much easier to remove as it is already loosened. Generally it is not necessary to wash potatoes as the soil is removed during the peeling operation.

Peeling

Tubers may be peeled in batch machines or in continuous machines. They generally are peeled using carborundum rolls or with brushes depending on the age of the tuber. As tubers age, that is, tubers in storage, the peel or periderm thickens up to several cells thick and it should be removed. However, freshly dug tubers and, in particular, immature tubers need only a light brushing to remove any peel. Potatoes may be peeled with the aid of caustic soda or lye.

Potatoes may be peeled with steam in steam pressure chambers. The dwell time in the steam should only be long enough

FIGURE 7.3 — "Waterless" Peeler/Scrubber
(Courtesy Kusel Equipment Co.)

to heat through the peel. Immediately after heating they should be cooled quickly to stop the heat penetration and the peel is then brushed away. Losses may be excessive if not operated properly.

With lye peelers, the potato are propelled through the lye by helical conveyors. The lye solution may be between 180–220°F (82–105°C), with a caustic solution between 5 and 15%. The dwell time may be as little as 10 seconds and as long as 1 minute depending primarily upon the age of the tuber. A new method of lye peeling uses little or no water. Immediately following immersion in the lye solution, the tubers should emerge directly into a cold water rinse to remove the lye and stop the digestion of the potato by the caustic soda. Lye peeling is a good method to remove the peel even in and around the eye of the tuber. However, lye is a caustic solution and care should be exercised in using it. It, also, may prevent a disposal problem. When using lye for peeling one must always maintain the concentration of lye to obtain uniform results.

The advantage of steam and lye peeling is that they both penetrate the whole area as well as defective areas and they cover all the potato versus abrasive peel which just grinds down the peel and into the flesh of the tuber if not operated properly.

Testing for the concentration is quite simple requiring only a standard hydrochloric acid solution of 0.1 N Hydrochloric acid and 1% solution of phenolphthalein indicator. A 10 ml sample of the lye is placed into a 250-ml volumetric flask and the flask filled with water and thoroughly mixed. A 10-ml solution is removed from the flask and placed into an Erlenmeyer flask with 100 ml of distilled water and 5 drops of phenolphthalein indicator added. This is mixed and then titrated with the 0.1 N hydrochloric acid. The number of ml of hydrochloric acid added to reach an end point is equivalent to the percent of caustic soda in the lye bath.

With the high pressure steam peelers, the potatoes are conveyed to the steam chamber where the steam cooks the potato immediately beneath the skin. The pressure is released and the potatoes are given a wash similar to those washed in the caustic solution to remove the loosened skin.

Peel losses can exceed 20% of the tuber weight depending on size of tubers, the age of the tubers, and peeling method, that is, dwell time in the peeler, use of caustic, and type and style of peeler. Peelers are designed for a given throughput and it behooves the operator to run the peeler at the right capacity for efficiency in peeling. Excess peel removal means lower yield and excess waste to dispose.

Washing, Sorting, and Trimming

These unit operations are important to produce a clean potato free from defects before slicing. Defects in potatoes may be prevalent due to diseased or bruised tubers. These defects are unsightly and cost the processor money and time to remove. For example a 1/2-inch defective area represents some 8 or 9 potato chip slices that may be defective. It is much easier and more efficient to remove the defective area ahead of the slicer and, of course, ahead of the fryer. With small size potatoes, it may be cheaper to eliminate the whole potato rather than try and trim the defect out.

FIGURE 7.4 — Potato Slice Speed Washer
(Courtesy Heat and Control, Inc.)

Some manufacturers now use electronic sorters ahead of the slicer to remove or kick out off-quality potatoes. By proper removal of defective areas ahead of the cooker or fryer one saves on energy to fry or dry the potato, saves on oil, and should increase the efficiency of the operation.

CHAPTER 8

Potato Preservation

There are four basic methods of preserving potatoes: Canning, Chipping, Drying and Freezing. Each of these will be discussed separately, however, emphasis will be placed on chip manufacture and Freezing, primarily Frozen French Fries. There are many similarities other than the actual preservation method. They do have wide differences in the finished product, particularly the shelf life. Potato chips have a normal shelf life of 10 to 12 weeks, but this can be extended through the use of gas packaging for another 3 to 5 weeks. Canned potatoes have a shelf life of some 30 months, however, salad canned potatoes have a shelf life of 24 months. The difference is due to the added acid (vinegar) in the formula. Frozen potatoes have a shelf life of up to 24 months, while dehydrated potatoes may have a shelf life of several years depending on how they are packaged and stored.

There are, also, differences in the convenience of the different products with potato chips being the most convenient. Here, it is a matter of opening the container and eating the product. In most cases the user may want to dip the chips and that means opening two containers.

The canned product is, also, a very convenient product. The canned potatoes may or may not need cooking or at least heating before serving, however, the potato salad can be eaten directly from the can.

Frozen potato products must be thawed and heated before serving. This is only a matter of minutes, but this requires specific equipment, that is, an oven, a microwave, or a pan or skillet.

Dried potatoes require reconstitution and this may take a few minutes to up to 20 minutes for most products.

Potatoes are sliced and fried in oil, lightly salted, bagged and sold as chips. The slice thickness varies from 0.050 to 0.080 and they may slice as flat chips or as wavy or ridged chips. They may, also, be partially dried before frying or dehydrated and sold as dried chips.

Today nearly two-thirds of the potato crop is processed with over 40% of the crop manufactured into frozen products, 18% manufactured into chips and some 16% dehydrated. Canned products only represent a little over 2%.

Regardless, all processed items are convenient and there is no waste.

CHAPTER 9

Canned Products

This is a relatively new processed product and its advantage is the ready convenience for the user.

Generally canned potatoes are an ideal outlet for small and low specific gravity potatoes (< than 1.070 specific gravity). Also, this is an ideal outlet for small sized potatoes, that is, from 1 inch or less to over $1^1/_{12}$ inches.

The canned potatoes may be packed in various styles, such as, whole, sliced, diced or Julienne style or as pieces. Whatever the style, they should be uniform within the container.

Specific potato varieties may be used, however, more often than not, early maturing tubers make an excellent source of potatoes. The variety or cultivar selected should not slough or breakdown when heated. The specific gravity should be below 1.070 and the potatoes should be packed in brine with calcium chloride added to keep the tubers from sloughing or breaking down during the canning process.

The potatoes may be left whole or cut for the style being used using sharp knives to assure clean cuts. The potatoes should be handled rapidly to prevent discoloration. They may or may not be washed after cutting to remove loose starch, slivers, and off size cuts. They should pass before inspectors to remove any defective pieces should they be present. If they have to be held before canning, they should be kept under water to prevent discoloration.

The potatoes should be filled into enamel lined cans or jars and covered with brine solutions or with hot water (190–200°F) (88–93°C) and salt tablets should be added. Generally salt is added at the rate of $1^1/_2$% by weight. Calcium salts (calcium chloride, calcium sulfate, calcium citrate, monocalcium phosphate or any mixture of two or more of these salts) may be added up to 0.051%.

The cans or jars should be steam vacuum closed or sealed, coded and immediately processed following a low acid canned foods process. For example, using a 303 size can and an initial temperature of 140°F (60°C), a cook time of 30 minutes at 240°F (115.6°C) is required. All process times and temperatures should be filed with FDA and developed with a processing authority.

Following cooking the canned product should be cooled to 95°F (35°C) before being cased and warehoused for shipment.

Potato Salad, a value added product, may, also, be formulated: (75 lbs. of 1/2 inch diced potatoes, 3 lbs. of chopped bacon, 1 1/2 lbs. of chopped onions and 7 ounces of fresh parsley) with a cover sauce made with 2.5 lbs. of flour, 2.5 lbs. of bacon fat, 5 lbs. of sugar, 1.5 lb. of salt, 0.7 ounces of pepper, 3.2 lbs. of 100 grain vinegar all made up with water to 5 gallons. This potato salad is canned similarly as to canned potatoes except that being an acidified product to a pH below 4.5, the process is somewhat different depending on the formulation, that is, the consistency and the pH of the product. Again the process is based on the recommendation of a processing authority, however, since this is an acidified product it can be cooked at a much shorter time or even a boiling water process may be used.

Yields will vary, but should equal 60 cases (basis of 303's) per ton of potatoes.

CHAPTER 10

Potato Chip Manufacture

Slicing

Without doubt this is the most important operation in a chip plant. Efficiency in slicing produces clean slices with no feathered edges and no torn slices. These kind of slices adsorb less oil and do not leave potato pieces in the oil to cause it to break down more quickly. The Urshel Company developed the Model CC Slicer for the chip trade. It is a centrifugal type slicer with the product entering the rotating impeller and the product is forced against an inner surface of the slicing head assembly which consists of 8 separate slicing head shoes and knives. As the product passes each knife in a smooth and uninterrupted manner, a slice is produced. The trailing edge of each slicing casting has a hardened stainless steel insert strip that is fluted to permit small bits of foreign matter to drop out ahead of the next knife.

FIGURE 10.1 — Slicer Head Assembly for the Chip Industry (Courtesy Urschel Laboratories)

When slicing at 0.060" (1.5mm) thick, the top capacity is 15,000 pounds per hour, however to operate at peak efficiency in terms of slice uniformity, the slicer should not be used for more than 8000 pounds per hour. When making flat slices the manufacturer states that 80% of the slices will not vary more than 0.004". By changing the blades, the cuts may be made wavy, crinkle cut or a "v" cut. Regardless of the type of cut, the knives must be changed as often as necessary to produce a good clean cut. They may have to be changed hourly or less if the cut is not a clean cut.

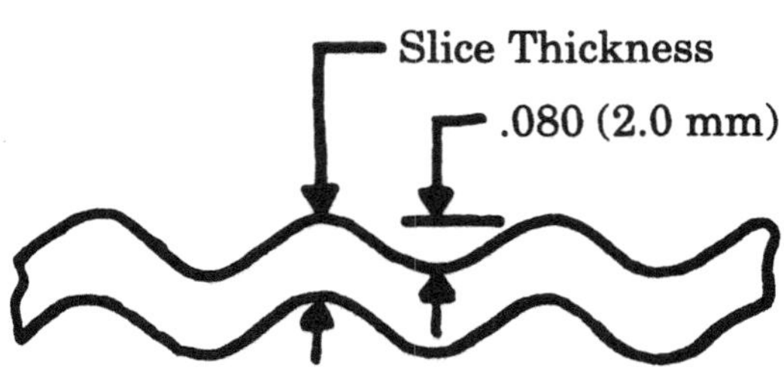

FIGURE 10.2 — Crinkle Slice

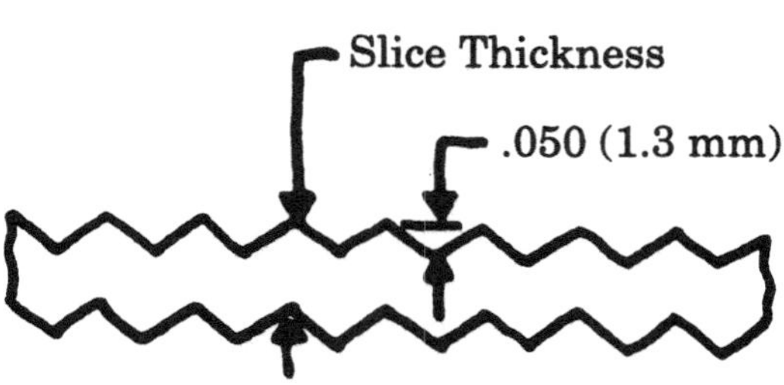

FIGURE 10.3 — V-cut Slice

High specific gravity and/or dirty potatoes may be problems for the slicer blades and the operator must test the slices regularly for uniformity of slice thickness and the type of cut being performed. Poor operation of the slicer will lead to many problems in the finished chip including high oil content, quicker breakdown of the oil, filter problems, etc.

As shown in Figure 10.4, a thickness gauge is used to measure the uniformity of the individual slice. If they are not uniform, the slicer should be evaluated to make certain that the blades are properly adjusted.

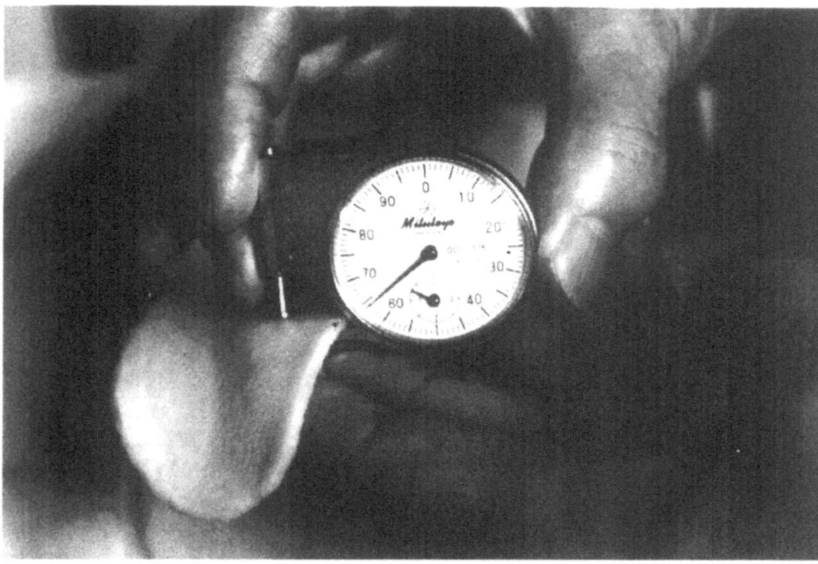

FIGURE 10.4 — Thickness Gauge

In addition to uniformity of the individual slice, I like to use number of slices per inch for control purposes. (See Figure 10.5). Slices per inch is a good measure of the cutting operation efficiency.

Washing Potato Slices

There are two schools of thought in regards washing following slicing. Some say no washing of the slices and others pay great attention to this operation. My experience indicates that the slices should be free of loose starch before entering the fryer to prevent oil breakdown and dark specks on the chips. I suggest washing by agitation and by counterflow of the water so that clean water is the last water on the slices.

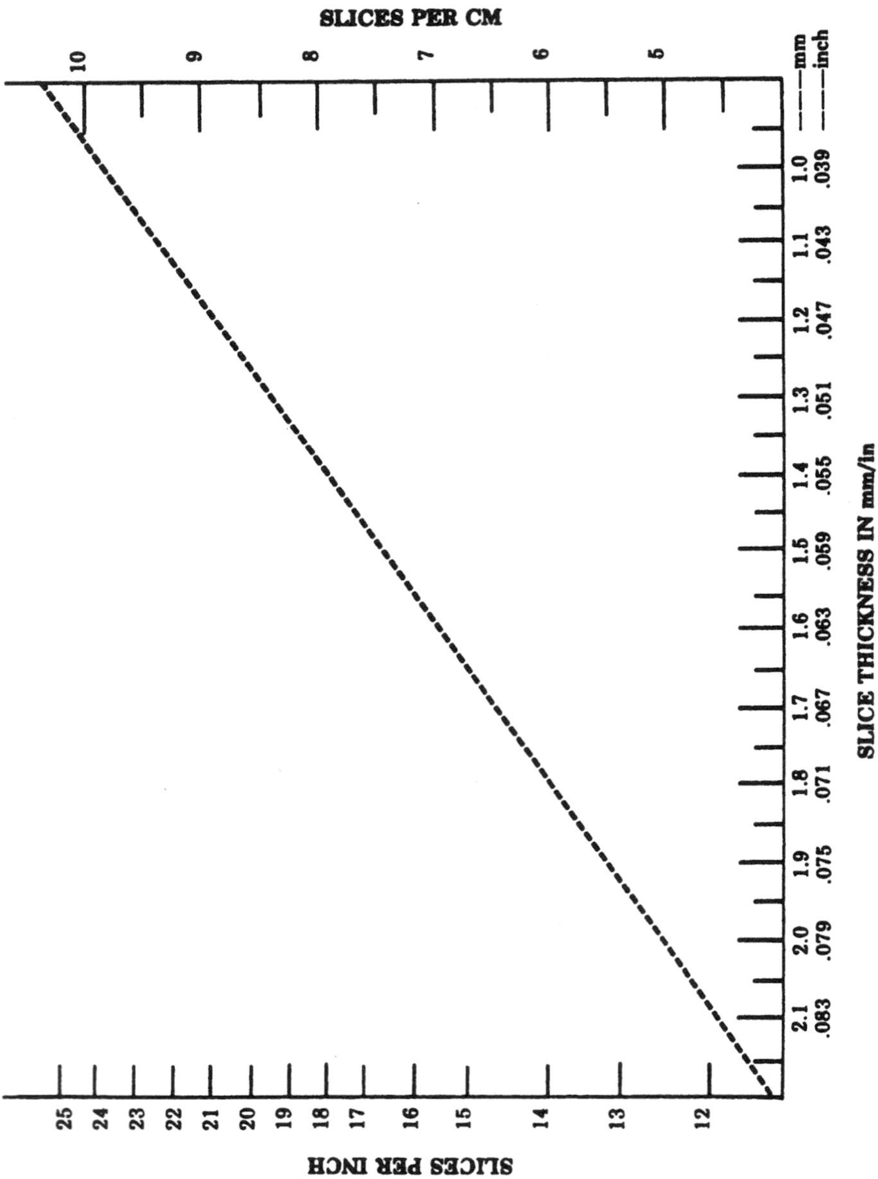

FIGURE 10.5 — Relationship of slices per inch or per centimeter to slice thicknesses.

The temperature of the water may be elevated if sugar is present in the slices. By taking the temperature up to 180°F (82°C) with a dwell time of 30 seconds one can improve the chip color 1 point on a 5 point color scale. However, the chip will taste more like a cooked potato and have a firmer texture. Some prefer the flavor and texture of the hot water treated slices to straight cold water wash.

Some people have difficulties in washing in that the water may be too soft, with chip slices lacking in texture. The water should be hard, that is, 250–350 ppm hardness and it should be at or near the pH of the tuber slices for greatest efficiency, that is, a pH of 6.2.

Prior to the slices entering the fryer, the slices may be blown dry or nearly dry with air knives to remove excess water (see Figure 10.6). The knives may be above and below the slices to remove maximum water before cooking. The slices should be shaken or vibrated to remove as much moisture as possible before entering the fryer Some manufacturers actually partially dry the slices before frying them.

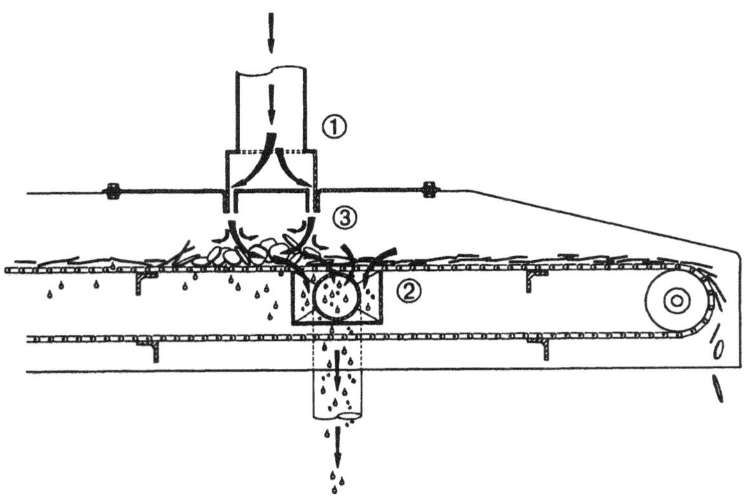

FIGURE 10.6 — Air Sweep Water Removal System

Drying and/or Frying

There are two basic methods of frying potato chips, that is,, the batch method or the continuous method. The batch method is used for the production of potato chips which are marketed as 'home style', 'old fashioned', or 'kettle style' potato chips. In the batch method the chips have a firm texture and some say a good bite. The operator rakes the chips to attempt to keep them in the oil while frying. The batch system requires more time up to 8 to 10 minutes per batch, but the temperature is usually lower and the oil content may be slightly higher.

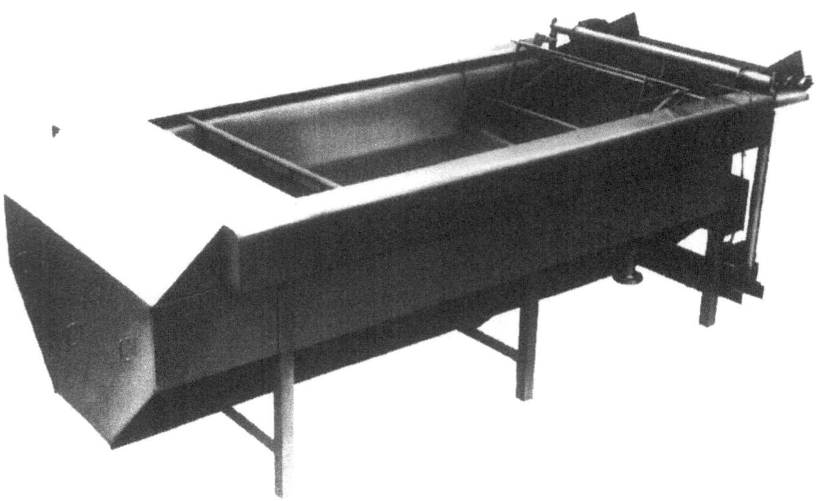

FIGURE 10.7 — Batch Fryer
(Courtesy Heat and Control, Inc.)

The continuous system is used by most manufacturers with widely different heating systems. some use direct fired while others use side heat exchangers. Further, some have converted their fryers to multi zone heating, that is, they can vary the heat in different sections of the fryer. This is good in that there are times this is the only way to control chip color. Multi-zone fryers are much more versatile, but do require good knowledge of frying temperatures in each zone.

FIGURE 10.8 — Potato Chip Fryer and Catch Box
with Main Oil Dump Set
(Courtesy Heat and Control, Inc.)

This unit operation is changing in that the oil content of the finished chip is of real significance to some customers. Today the market share of low fat/no fat chips is increasing and is up to 11 %. I happen to be one that thinks potato chips should be fried in oil for best flavor. However, drying by itself or partial drying prior to frying is being used. Drying takes much more time and the resultant product is very mealy with little or no flavor other than dried potato flavor. It never has been conducive to high potato consumption and having lived on dried potatoes while in the military, I do not want to go back to dried potatoes. Regardless, some in the industry think this is the way to make potato "chips". This method does not conform to the Saratoga style chips, the style that this industry built its reputation on. Nevertheless, oil content of chips should be and perhaps will have to be much better controlled.

Plimpton in her dissertation work used Norchip tubers sorted into two given specific gravity lots and developed the following equation:

$$y \text{ (oil content)} = 2543.63 - 198.2 \text{ (specific gravity)}$$

The equation is slightly different than Lulai and Orr indicating that the oil contents were lower until a specific gravity greater than 1.100 was obtained. Further, the higher oil content in Lulai and Orr's work could be attributed to the slice thickness as they only used 0.050" slices. Plimpton's studies included 3 slice thickness, that is, 0.050", 0.060" and 0.070". The data are shown in Figure 10.9. In the OSU studies, specific gravity differences accounted for up to 43% of the variation in the oil content when all factors are constant and when frying to a moisture content of 1.5%.

There are other ways to control the oil content of potato chips and they should be exploited, as many in the industry are doing. Generally the chip manufacturer controls the oil content of his chips by (1) careful selection of variety/cultivar that are high in

Figure 10.9 — Frying Time in Seconds to 1.50% Moisture Content for Flat Style Potato Chips for Given Specific Gravities at Three Slice Thicknesses and at Four Frying Inlet Temperatures

Specific Gravity	Slice Thickness	Inlet Frying Temperatures (°F) 375°	350°	325°	300°
1.065	0.070	154	209	261	316
	0.060	101	146	212	300
	0.050	53	111	165	233
1.075	0.070	150	200	257	308
	0.060	96	141	212	292
	0.050	49	104	160	228
1.085	0.070	145	196	252	300
	0.060	96	137	207	285
	0.050	51	100	156	225
1.095	0.070	143	187	245	296
	0.060	91	131	198	277
	0.050	45	96	153	220
1.105	0.070	141	178	230	285
	0.060	86	128	194	270
	0.050	40	93	149	218

specific gravity; (2) slicing thick with less potato surface area for oil adsorption and (3) high temperature with shorter dwell time in the fryer.

The second key variable to control the oil content is the surface area of the slice, that is, the slice thickness as thin slices will adsorb more oil since slice thickness is a surface phenomena. The data in Figure 10.10 clearly show the effect of slice thickness at given inlet frying temperatures on oil content.

Oil content can vary from 0 up to more than 40%. In my opinion,

FIGURE 10.10 — Effect of Slice Thickness on Oil Content of Potato Chips Fried at Various Temperatures to a Moisture Content of 1.50% for Regular Style Chips

Slice Thickness		Oil Content in % for Frying Temperatures		
Inches	Ct/In.	375°F (190°C)	350°F (177.6°C)	325°F (163°C)
0.050	20	41.3	42.1	42.9
0.060	16	36.3	37.7	39.2
0.070	13	33.8	35.9	38.0

the oil content should be between 26 and 30% for ideal chips. The data in the Figure 10.9 shows fry time in seconds for various specific gravity tubers at given slice thickness and at 4 different inlet fry temperatures for flat style chips.

Frying in oil is basically a drying process, complete or partial replacement of frying in oil with an alternative drying operation would decrease the oil content of the finished chips by decreasing the time available for oil adsorption. Vacuum fried chips generally have less oil than conventional fried chips.

Mancuso and Capossela showed that dipping raw slices in hydrogenated oil at 100°F (37.8°C) for 5, 10, and 15 minutes succeeded by drying in an air dryer for 10 minutes at either 315°F (157°C) or 325°F (163°C). Potato chips produced in this manner were lower in oil with the shorter soak period yielding chips with a 23.8 % oil content followed by 24.9 and 27.3% for the 10 and 15 minute oil soak periods.

Nack and Shutz patented a method of utilizing a fluidized bed of hot sodium chloride particles to cook the raw potato slices to their final moisture content. Oil then could be applied by spraying permitting control of the amount of oil in the final product.

Goulston received a patent for a solvent extraction process in which n-hexane, alcohol's such as ethyl alcohol, ketones such as acetone, pure liquid hydrocarbons, chlorinated hydrocarbons and low boiling petroleum fractions could be used to extract as much as 43% of the oil from the chips.

Difluorodichloromethane (R-12) has been used as a solvent for oil extraction from fried foods. This is reaidly available, recoverable, and cleared by FDA. As much as 33% of the oil in potato chips can be extracted with R-12.

Hannigan reported on a patented method in which chips fried by conventional processes were extracted with supercritical carbon dioxide removing over 50% of the oil in potato chips. This extraction procedure solubilizes fats and oils without removing proteins and carbohydrates and maintains a good texture and flavor characteristic of potato chips.

Baroudi fried Norchip potatoes at 375°F (190°C) for 12 seconds followed by drying on a continuous belt oven for 3½ to 4 minutes (temperature in the dryer was 220°F (104°C), falling to 180°F (82.2°C) after 27 seconds, and rising linearly to 310°F (154°C) after 180 seconds) and finished frying at 290°F (143°C) for 65 seconds, yielding potato chips with 33% lower oil content than the check sample at 46% oil content.

Thus, one can conclude after several attempts to control or reduce the oil content in potato chips, there are no simple solutions. However, partial drying before frying appears to be a good procedure to help control the oil content of potato chips.

Another important factor to keep in mind when frying potato chips and that is the "turn-over" of the oil. Turn-over simply put means the amount of oil that is used during a given run. Ideally, one should hope to turn the oil each shift. Assuming one has a 1,000 lb. fryer, that is, the fryer holds 1,000 pounds of oil. And assuming that the oil content in the finished chip equals 33%. Then, for every 3,000 lbs. of chips, the oil would have to be totally replaced. Obviously, in practice the oil is gradually replaced during a run and, in this example, one would have to replace all the oil during the manufacture of 3,000 lbs. of chips. By constantly turning the oil over, it should always be fresh and properly flavor the chips.

FIGURE10.11 — Oil Tanker Delivery Frying Oil To A Chip Plant

Much could be said about the kinds of oil in use today from corn, soybean, cottonseed, peanut, palm, etc. including the new patented Olean. Each type of oil produces different flavors to the chips and some manufacturers specify specific oils as the market they have developed demands for that type of oil flavor. Some manufacturers always use blends to keep the flavor more neutral and to better compete in the oil market place.

Dried potato chips are dried in conventional ovens with no oil. The practice of drying takes much longer, but this practice may be easier to control the color and moisture content of the chips. The thinner the slice the quicker the drying, however, most dried chips should be sliced thicker than fried chips. Dried chips will develop off flavors quicker than fried chips unless they are blanched before drying. Hot water blanch is acceptable just before drying. As we learned during World War II, oxygen scavengers should, also, be used in packing of dried potatoes to better control any off-flavors.

FIGURE 10.12 — Workers inspect chips as they come off the fry line.

Inspection

Chips are inspected following the frying/drying operation to remove discolored, burned, or defective chips. In the past this was accomplished with "chip pickers", that is, people on the line to pick out the bad chips. Today, electronic equipment is available to "kick out" the defects, the off colored or the unsuitable chips. The advantage of the electronic equipment is that it works all the time and is as effective as management wants it to be by proper adjustments. It should be noted that any pick-out, kick-out, or throw out is expensive at this juncture in a chip plant and it can be costly. The chip may be worth some 25 cents per ounce here, whereas the raw material was worth only 3 to 4 cents per pound. One need only compare and see the quick loss of profit by improper operation of this unit operation. Defective chips are not desirable and it behooves the industry to present clean defective chips to the customer.

Salting and Seasoning

The taste of any snack food is the single most important aspect of the finished product. The key is to give the customer what they expect every time. Most snacks including potato chips are

FIGURE 10.13 — Tumble Drum Seasoner
(Courtesy Heat and Control, Inc.)

surface salted. Salt is the least costly of all the ingredients used in making potato chips and it is used to enhance the flavor the chips. Generally chips are salted direct from the fryer to take advantage of the hot oil to bind the salt crystals to the chip. The Snack Food Association has recommend a salt level of 1.75% plus or minus 0.25%.

Potato chips may be seasoned with BBQ, Vinegar and Onion, Cheese, and other flavors. At the present time about 25% of the chips are flavored with one or another type of season. Generally, the seasoning is applied as a topical season in rotary drums or sprinkled on top like salt is applied. Of course, many chips are consumed with given dips and salsas. Usually dip chips are cut thicker such as the wavy style chips.

Cooling

Potato chips are usually cooled prior to packing. The cooling eliminates moisture in the bag and improves the texture of the chips. Cooling, if done with holding bins, may provide a surge area for holding chips. However, one should always remember, the more times chips are dumped, handled, conveyed, etc. the greater the breakage and the more hash' in the final bag or container.

FIGURE 10.14 — Chips Cooling On A Conveyor

Packaging and Casing

Most snacks are packaged on form, fill and seal machines using specific bags for given regions of the World. The metallized foil is a prominent type of package. The selection of the quality of the package should be based on the contemplated shelf life of the product. Long shelf life (over 10 weeks) require low moisture/vapor proof films. All packages should be properly filled, sealed, and coded. The bag should be full and may be gassed packed using an inert gas (nitrogen) to better keep the quality of the chips while in the marketing channels.

The declared label weight must be met on potato chips. This is not a big problem when using automatic weigh scales that portion out the product as needed. Check weighers are an important part of this industry today.

All packaged snacks should be properly coded with pull date or last date of use included. The code is to protect the manufacturer as well as the customer and the more detailed information may prove very helpful. I have seen operations wherein every case is consecutively coded so that one knows exactly when manufactured.

The last check on the operation is by the case packer to make certain that the bag is properly sealed and packaged. Chips are cased to improve and facilitate handling practices. Nevertheless, they should be handled with tender loving care to prevent breakage and retain product quality.

Warehousing

Packaged chips are placed in warehouses awaiting shipment to distribution centers or stores. Warehoused chips should be handled with care and the practice of First In, First Out (FIFO) should be followed. Warehouses should be operated at or below room temperatures with low humidities for holding of the chips to increase shelf life.

FIGURE 10.15 — Unit Operations For Chip Manufacture

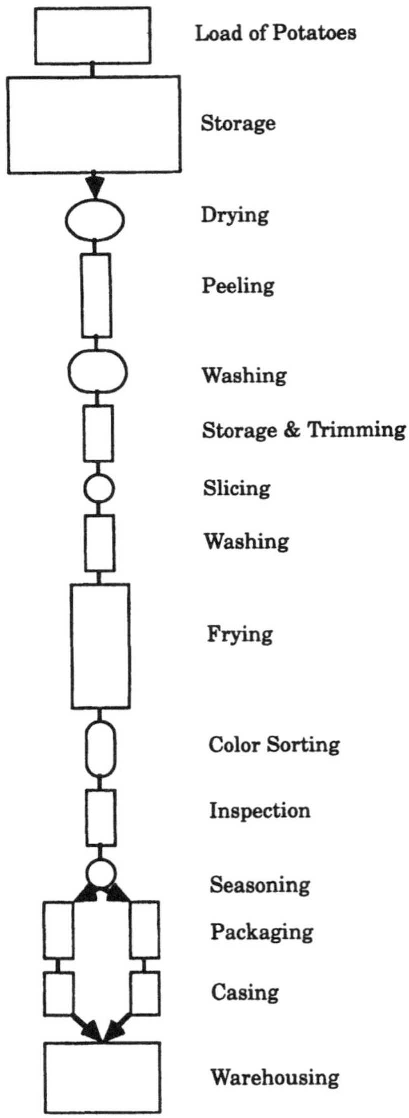

CHAPTER 11

Dried Potato Products

Dried potatoes have been around since the beginning of our knowledge about potatoes. However, only in recent years have the products been of high quality. The experiences during World War II changed the thinking on how to properly dry potatoes. And in the 1950's, the USDA came to the forefront and developed methods for the manufacture of new style potato flakes, one of the better products on the market today.

Potatoes can be dried in diced form, strips or Julienne, riced and manufactured into flakes. They are prepared as previously described.

The raw potatoes are diced into 3/8 x 3/8 inch pieces, blanched in hot water to inactivate the enzymes and then spread on trays or onto the dryer belt and carefully dried to prevent case hardening of the product. Case hardening is caused by too fast a drying curve and not allowing the internal moisture to diffuse to the surface. They are dried to a moisture content of 3–4 %, gas packaged using an oxygen scavenger such as ascorbic acid, sealed until used. Diced potatoes are used for soups, home fries, salads and mashed potatoes. The product can be kept for several years, if done correctly. One difficulty with the product is that it is bulky even though many efforts have been made to compress the dices. Compression of diced potatoes can be done, if done while the product is still hot coming from the dryer, but they never reconstitute as well as uncompressed diced potatoes.

Strips or Julienne potatoes are dryed similarly to the diced product. The strips are cut 3/8" and vary in length, depending on the length of the potato. The product has too many fines and wasted cuts that make it difficult to become an efficient method of potato preservation.

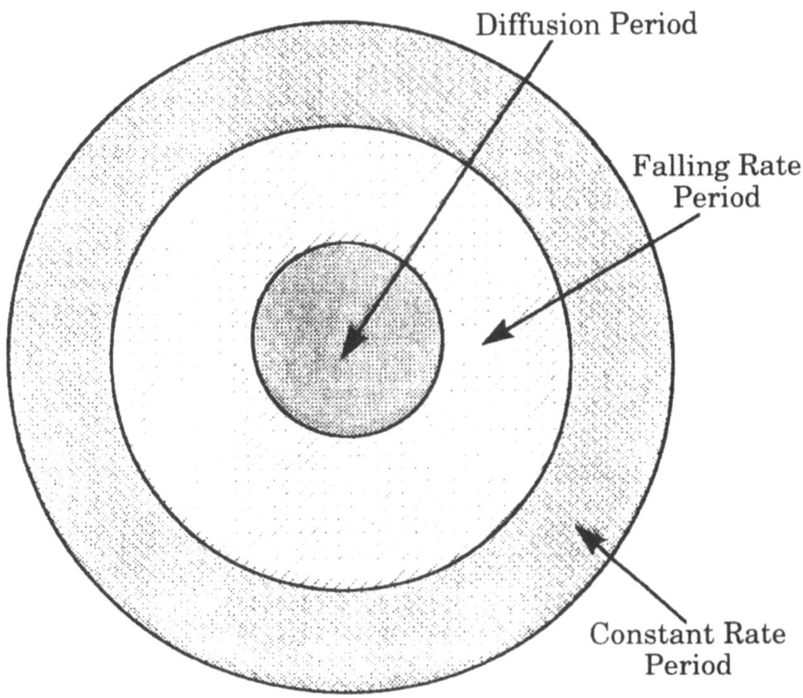

FIGURE 11.1 — Moisture Removal

Mashed potatoes were long the favorite method of preserving potato. The product is thoroughly cooked, and then extruded or riced onto drying trays or belts. It is quite easy to dry if the tray or belt load is uniform as the time and temperature can be lower than for drying diced or Julienne potatoes. The finished dry product with a moisture content of less than 3% can be kept for several years if properly packaged. However, the method is being replaced by the flake process in that a wider range of total solids content potatoes can be used.

The potato flake process is a quick method of drying mashed potatoes using a drum dryer. It is an improved method of retaining the potato flavor. It is a very simple process in terms of machinery, time, and affect on product quality. The process relies on cooking the raw potatoes in water at a temperature of about

FIGURE 11.2 — Drying Curve vs Conveyor Drying Zoning

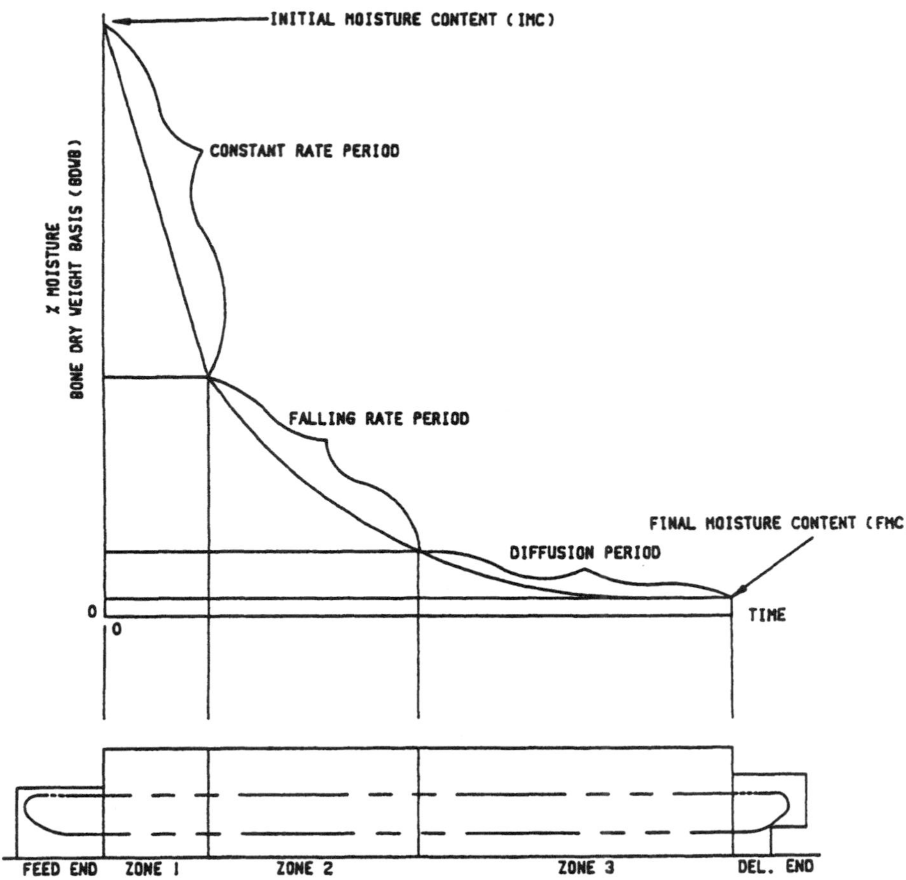

165°F (74°C) for about 20 minutes before final cooking of the potatoes in steam. The advantage is that the potatoes absorb more water during reconstitution. Precooking is believed to swell the potato cells to distend them and to separate them more readily during mashing. Further, lower solids content potatoes can be used successfully. The mashed product is drum dried and then broken into flakes prior to packaging. It rehydrates quickly and is widley used by the HRI trade because it can be prepared quickly and there is no waste.

Potatoes are also dried as slices that are cut 1/8" thick for use in Au Graten style potatoes. They can be seasoned and cheese flavored to meet many styles of use.

CHAPTER 12

Frozen French Fries and Other Frozen Potato Products

Without question this is the fastest growing segment of the processed potato industry and mostly due to frozen French fries. The fries are primarily manufactured by firms in the Eastern USA, Idaho and Washington.

There are many styles of Frozen French fried potatoes:

Straight Cut, that is smooth surface as in home fries (varies from $^3/_8$" to $^5/_{16}$").

Crinkle Cut, same as above but with corrugated cut.

Strips, that is elongated pieces with practically parallel sides and of any cross-sectional shape, for example $^1/_4$ x $^1/_4$", $^3/_8$ x $^3/_8$", $^1/_2$ x $^1/_4$", or $^3/_8$ x $^3/_4$".

Shoestring, that is, strips either straight or crinkle cut with a cross section area predominantly less than that of a square measuring $^3/_8$ x $^3/_8$".

Slices, that is, pieces of potato with two practically parallel sides and which otherwise conform generally to the shape of the potato.

Dices, that is cubes.

Rissole, that is, small whole or nearly whole potatoes.

Length designations vary, but the following are the terms in use today:

Extra Long, 80% or more are 2" or longer, and 30% or more are 3" or longer.

Long, 70% or more are 2" or longer, and 15% or more are 3" or longer.

Medium, 50% or more are 2" or longer

Short, less than 50% are 2" in length or longer.

Some have believed that the variety of the potato was the secret to success. However, many different varieties are used, but the first priority should be high specific gravity. Varieties such as the Russet, Kennebec, Shepardy are widely used. A second priority should be large size for long cuts. Large size generally means lower peel loss and cuts suitable for the HRI (Hotel, Resturant, Institutional) trade. The processor procures the raw material based on his or her experience and the availability of suitable potatoes in the area. Contracting is a basic requirement for success in this industry.

Today the French fries may be cut using water guns with stationary knives cutting them lengthwise. The fries may be sorted after cutting to remove slivers, nubbins and any defective units. Electric/electronic eyes are becoming quite popular for sorting for color and defects.

The potatoes are handled as for potato chips, but the processing differs in a number of areas.

Following cutting and washing, the fries go thru a series of shakers to remove as much water as possible before they go to the blancher. Blanching is an important unit operation for both dried and frozen products. Blanching serves to inactivate the enzymes that could cause the product to be off-flavored, discolored and/or changes in the texture; blanching helps to stabilize the color and retain the texture of the product; and blanching will reduce fat or oil absorption by gelatinizing the surface layer of starch. Blanching may, also, be used to reduce the sugars, that is, the leaching of the sugars from the cut and peeled potato. Blanching reduces frying time since the potato is partially cooked in the blanching operation.

If the tubers have high sugar content the hot water blanch will reduce some of the sugars by leaching. Generally two blanchers are used with a lower temperature in the first blancher at 165°F (75°C), and at 195°F (90°C) in the second blancher for texture effects. The product is cooled in between blanching to improve the texture of the final product. The time and temperature of the blanch is determined by the total inactivation of the enzymes. The blanched product is evaluated for enzyme activity, using the peroxidase test.

The blanched product should be cooled and as much water as possible removed by shaking or blowing the water off the product and/or partially dried before being fried.

FIGURE 12.1 — French Fry Fryer
(Courtesy Heat and Control, Inc.)

The frying is a partial frying for the HRI trade and a more complete frying for the consumer/retail market. Next, the fries go to a freezing tunnel and then they are filled and packaged.

There are many styles of frozen potato products and many carry trade marks as they were developed by Willard or other firms. The different styles of processing for some products are shown in the flow Chart from the USDA publication ARS 74 8.

One recent innovation is the use of additives (starchbased coatings) and clear coatings that are nearly or invisible to the naked eye to the finished product to improve texture, improve flavor, and improve the outward appearance of the fries.

FIGURE 12.2 — Partial Flow Sheets for Processing Frozen Potato Co-products

Patties	Mashed or Whipped	Diced	Puffs	Hash-brown	Dehydro-frozen Pcs.
Blanching	Cooking	Dicing	Cooking	Cooking	Slicing
Dewatering	Mashing	Blanching	Mashing	Cooling	Blanching
Cooling	Mixing[2]	Dewatering	Mixing[3]	Shredding	Dewatering
Shredding	Finishing	Cooling	Extruding	Inspecting	Dehydrating[4] (Partial)
Mixing[1]	Whipping	Inspecting	Deep fat frying	Packaging	Cooling
Shaping	Cooling	Freezing	Cooling	Freezing	Inspecting
Inspecting	Inspecting	Packaging	Inspecting	↓	Freezing
Packaging	Packaging	↓	Freezing	↓	Packaging
Freezing	Freezing	↓	Packaging	↓	↓

[1] Mixed with flour and seasoning.
[2] Mixed with milk solids and salt.
[3] Mixed with flour, egg solids, shortening and seasoning.
[4] Dehydrated to 50% or more weight reduction

FIGURE 12.3 — Flow Chart (Courtesy Heat and Control, Inc.)

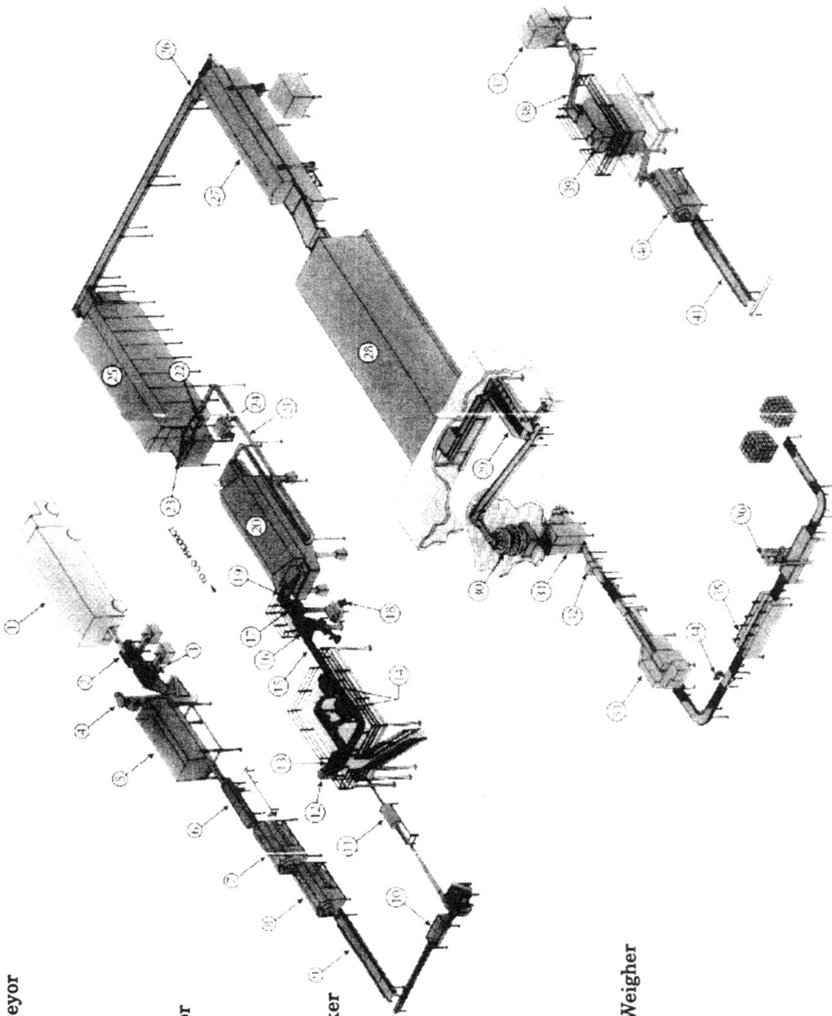

EQUIPMENT LIST

1. Truck w/Unloading Conveyor
2. Dirt Eliminator
3. Product Sizer
4. Destoner
5. Metering Bin
6. Inspection Conveyor
7. Abrasive Peeler
8. Washer
9. Inspection/Trim Conveyor
10. Vortex Tank
11. Water Knife
12. Decelerator
13. Dewatering Shaker
14. Cutters
15. Flume
16. Dewatering Grader/Shaker
17. Inspection Shaker
18. Tank w/Pump
19. Feed Shaker
20. Blancher
21. Flume
22. Dewatering Shaker
23. Feed Shaker
24. Tank w/Pump
25. Dryer
26. Feed Shaker
27. Fryer
28. Freeze Tunnel
29. Grader
30. Computer Combination Weigher
31. Form/Fill Machine
32. Metal Detector
33. Case Packer
34. Checkweigher
35. Case Settler
36. Case Sealer
37. Metering Bin
38. Feed Conveyor
39. Steam Peeler
40. Washer
41. Inspection/Trim Conveyor

CHAPTER 13

Technology & Quality Assurance

The technology of the potato and processed potato products can be very complicated. There are many areas to delve into, however, I prefer to look at the technology from an applied standpoint and discuss the basic areas as they relate to the quality of the finished products.

First, some brief statements relative Quality Assurance. More in depth coverage of this and the following topics, that is, Color, Texture, Absence of Defects, and Flavor are to be found in my book, *Total Quality Assurance*.

Quality assurance is a program designed to provide confidence for management and the ultimate user of the processed product by the customer. Quality assurance can be divided into three basic areas, that is, control, evaluation, and auditing of the food system or the products being manufactured and marketed. As stated in the above TQA book, "a firm is in business to produce a product intended for sale to a customer from which the firm hopes to make a profit. The key word is the customer. The customer is the one a firm must satisfy and it is the customer who ultimately establishes the level of quality employed by the manufacturer. The customer is management's guide to quality and this is what the firm must build its specifications and label requirements around. Only by having a planned program can a firm continue to succeed in supplying the customer with what they expect".

The control of quality is essential to success. Quality control means to regulate to some standard or specification. It is usually associated with the production line, that is, the specific unit operations and their processes. Quality control is a tool for the production worker to help him or her operate their unit operation in accordance with the predetermined parameters or

FIGURE 13.1 — Unit operations in potato chip manufacture with QA-suggested control and evaluation characteristics and attributes

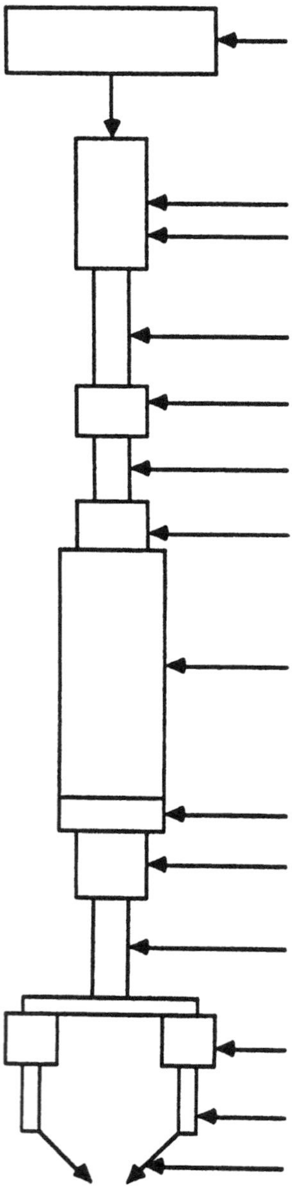

POTATOES by cultivars: maturity, specific gravity, freedom from defects, shape size, pulp temperature, count/10 lbs., reducing sugars, soil and HEM. Stored tubers: relative humidity, temperature, time of holding, condition of storage and sucrose, reducing sugar content.

DESTONING: Removal of stones, metal, sticks.

PEELING: dwell time, amount of water, amount of chemicals, if any, efficiency of operation, amount of waste, amount of wastewater.

INSPECTION & TRIMMING: removal of any defective area, manually or electronic sorting.

SLICING: uniformity and thickness.

WASHING to remove excess starch: hardness of water, temperature of water, dwell time.

DRAINING and/or AIR DRYING: removal of excess moisture and electronically sorting defective slices.

FRYING: dwell time and temperature—delta temperature varies with specific gravity and/or slice thickness. Oil quality—% free fatty acids, peroxides and sediment or clarity of oil, flavor of oil, oil absorption rate.

SALTING and/or SEASONING: Type & amount.

SORTING: Electronic color and defect sorting to remove from quality chips—% removal of defects and off-colored.

VISUAL INSPECTION: % removal of defects and off-colored, burned, etc.

PACKAGING: Integrity of seal, codes, net weights, % breakage, bulk density, amount of air/gases.

CASING & HANDLING

WAREHOUSE: time, temperature, RH, handling and FIFO.

specifications for any given quality level. Quality assurance personnel must spend a large portion of their time and talents in training the operator to work within the specifications and requirements set forth by management.

The primary objective of a quality assurance program is to obtain adequate information on all product factors or characteristics that affect the quality of the product. The intelligent interpretation of this information provides management with a quality index of the entire operation. The information serves as a management guide for the exact quality being processed from a given raw material or it may provide management with information necessary for the processing of the product to obtain a given quality. Therefore, quality assurance serves to:

1. Develop specifications for the control of raw materials being procured;
2. Establish parameters for the control and improvement of product quality;
3. Improvement of processing methods to lower processing costs, improve efficiency and increase productivity;
4. Increase order and compliance with Current Good Manufacturing Practices and improve food plant sanitation;
5. Remove any hazards in the product or process and control the critical control points to levels that comply with HACCP;
6. Maintain consumer confidence in the manufactured product and the firm.

For the success in any quality assurance program, there are several facets for success. These are fully discussed in *Total Quality Assurance*, so I will only note these as follows: (1) Organization—only responsible to top management and report directly to management, but share information with all other departments, (2) Obtain qualified personnel, (3) Understand sampling plans and know how to sample incoming materials, production lines, and finished products, (4) Understand specifications and standards of quality, that is, how to develop them and how to interpret same, (5) Know the measurement techniques and be able to demonstrate their use for line personnel, (6) Understand the techniques for process control and the statistical

FIGURE 13.2 — Daily Quality Assurance Record of Potato Chip Manufacture

	A	B	C	D	E	F
1	UNIT OPERATIONS/TIME		1-	2-	3-	4-
2	STORAGE	RH				
3		TEMPT				
4		TIME				
5		CONDITION OF T				
6		CONDITION OF S				
7						
8	DESTONING	% REMOVAL				
9						
10	PEELING	% REMOVAL				
11		# OF WASTE				
12						
13	WASHING	WATER HARD				
14		TEMPT OF WATER				
15		DWELL TIME				
16						
17	SORT & TRIM	% GR REMOVAL				
18		% HH REMOVAL				
19		% ROT REMOVAL				
20		NO OF PERSONS				
21						
22	SLICING	SLICE THICK				
23		NO/1 INCH				
24						
25	WASHING	HARDNESS				
26		TEMPT OF WATER				
27		DWELL TIME				
28		% WATER CARRYOVER				
29						
30	FRYING	INLET T.				
31		OUTLET T.				
32		DELTA T				
33		TIME				
34		% OIL DRIP				
35		% FFA				
36		% OIL ON CHIPS				

FIGURE 13.2 — Continued

	A	B	C	D	E	F
37		% MOISTURE				
38		TEXTURE				
39		FLAVOR OIL				
40		FLAVOR CHIP				
41		COLOR OF OIL				
42		COLOR OF CHIPS				
43						
44	COLOR SORT	WAVELENGTH				
45		% REMOVAL				
46						
47	INSPECTION	% REMOVAL				
48		NO PERSONS				
49						
50	SEASONING	TYPE				
51		% APPLIED				
52						
53	SALT	% APPLIED				
54		% ON CHIPS				
55						
56	PACKAGING	ACTUAL WT				
57		RANGE OF 10 PKGS.				
58		INEGRITY OF SEAL				
59		% BREAKAGE				
60		BULK DENSITY				
61		CODE				
62						
63	CASING	AMT. AIR BAG				
64		INTEGRITY OF SEALS				
65						
66	WAREHOUSE	TIME IN STORAGE				
67		CONDITION OF STORE				
68		HANDLING PRACT				
69		TEMPERATURE				
70		RELATIVE HUMIDITY				

interpretation of data and problem solving principles, and (7) Know the flow of the product through the factory (see WAGCO Model factory) and know the various unit operations including their parameters (see Daily Quality Assurance Record Form, Figure 13.2) and know the limits of quality attributes and characteristics for each level of quality (see Quality Levels for Potato Chips, Figure 13.3).

There are many other concerns with product quality control and evaluation and these will be covered in the following chapters:

1. Understanding the potato tuber;
2. Color and Color Measurements;
3. Texture and the control of Texture, that is, Moisture and Processing Parameters;
4. Flavor and Factors Affecting Flavor including cooking Oils, Salt and Seasonings;
5. Absence of Defects.

In addition to the above, the next two chapters are part of the technology of potatoes and potato processing to assure success in this industry:

6. Current Good Manufacturing Practices (CGMP) and Food Plant Sanitation;
7. Hazard Analysis and Critical Control Points (HACCP) including establishment of Critical Limits (CL).

A listing of some of the many terms and terminology will conclude this section of the book.

FIGURE 13.3 — Quality Levels for Potato Chips

Attributes or Characteristics	Quality Levels or Ratings				
	1-A Excellent	2-B Good	3-C Acceptable	4-D Poor	5-E Off
SFA color	1	2	3	4	5
Agtron M30 90/100	>65	54-64	44-53	33-43	<32
Color grade score	26-30	21-25	11-20	6-10	0-5
% Moisture	<1.50	1.51-2.00	2.01-2.50	2.51-3.00	>3.01
% Wet centers	<1	1-1.5	1.5-2.0	2.0-2.5	>2.5
% Oil content	<35	35-38	39-40	41-42	>42
Texture grade score	19-20	16-18	11-15	6-10	<5
% Broken chips/hash	<10	11-15	16-20	21-25	>25
% Minor defects (<1/4")	<5	6-10	11-15	16-20	>20
% Major defects (<1/2")	<3	3-5	6-8	9-12	>12
Absence of defect score	19-20	16-18	11-15	6-10	<5
% Salt	<1.5	1.6-1.75	1.76-2.0	2.1-2.5	>2.5
% Free fatty acid	<0.3	0.3-0.5	0.5-0.7	0.7-0.9	>0.9
Flavor score	26-30	21-25	11-20	6-10	<5
Total score points	>90	80-89	70-79	60-69	<59

FIGURE 13.4 — USDA Color Standards for Frozen French Fried Potatoes, 4th Ed., 1988 (Courtesy Kollmorgen Corp.)

CHAPTER 14

Understanding the Workings of the Potato

The organic chemistry of the potato is a basic and fascinating subject and of much concern to the user of the potato, although we do not know all that we should know to understand the workings of potato tubers. With the advent of sophisticated chemical methods, studies have been underway to determine the effects of the carbohydrate compounds, the organic acids and the amino acids on the color, texture, and flavor of potatoes and their processed products.

The potato is grown, handled, stored and processed under many varied conditions and one has to wonder why certain things happen the way they happen. To start with one must remember that the growing potato plant is a manufacturing factory. In simplest form carbon dioxide from the air and water from the ground are synthesized in the green plant with the aid of sunlight to make simple sugars. The process is called photosynthesis and it is reversible under many conditions including stress conditions. The process continues as long as the plant is not stressed from lack of water, lack of fertilizer nutrients, lack of sunlight, diseases or insect problems or excesses of nutrients, water etc. The plant has the ability to make simple sugars and synthesize them to sucrose, and sucrose is then translocated to the underground stem and further synthesized to starch and stored as such.

This whole photosynthetic process will continue as long as the plant is green and growing with a tremendous build-up of starch as the stored chemical. The main constituent of the tuber, of course, is water. The starch consists of two polymers, that is, amylose (21 to 25%) and amylopectin (75 to 79%). Both polymers consist of glucose with the amylose as a straight chain while amylopectin is a branched chain.

If the plant is stressed and/or if the tuber is harvested immature the sugars never are all converted to starch and the result is that sucrose will break down to simple reducing sugars rather than be synthesized to starch. These kind of reactions present serious problems to the user of the tubers. Thus, it behooves the grower to make certain that the plant is allowed to come to full maturity and that the vines die or are killed so that they no longer are capable of manufacturing and translocating sugars.

The mature tubers are then ready for digging and after proper suberizing they are can be stored as all the sugars should be translocated and stored as starch. Immature potatoes will have large amounts of sugars. They should never be stored for use by chip users. Immature potatoes can be manufactured directly into chips under most conditions because the reducing sugars from sucrose or from photosynthesis will respire and should be below 0.15%.

Tubers prior to harvest, during harvest, and after harvesting must never be subjected to temperatures below 50°F (10°C) as temperatures below this will trigger high rates of conversion of sucrose to reducing sugars, particularly fructose.

Without question the greatest difference in tuber composition is the varietal difference. The following two chromatographs show great differences in two of the acid components of two varieties, that is, Monona and Sebago. Alpha keto glutaric acid is extremely high in Sebago Vs very low in Monona. The same thing is true for the amounts of caffeic acid. "Fingerprinting", as shown in these chromatograms, of individual tubers might give us a clue as to why varieties react differently. Monona is a good chipping variety while Sebago is not a good processing variety. These chemical differences could explain some basic fundamentals of tubers.

One must, also, remember that the tuber is alive and does respire. The amount of respiration will be dependent on the handling practices and the storage temperatures and humidities.

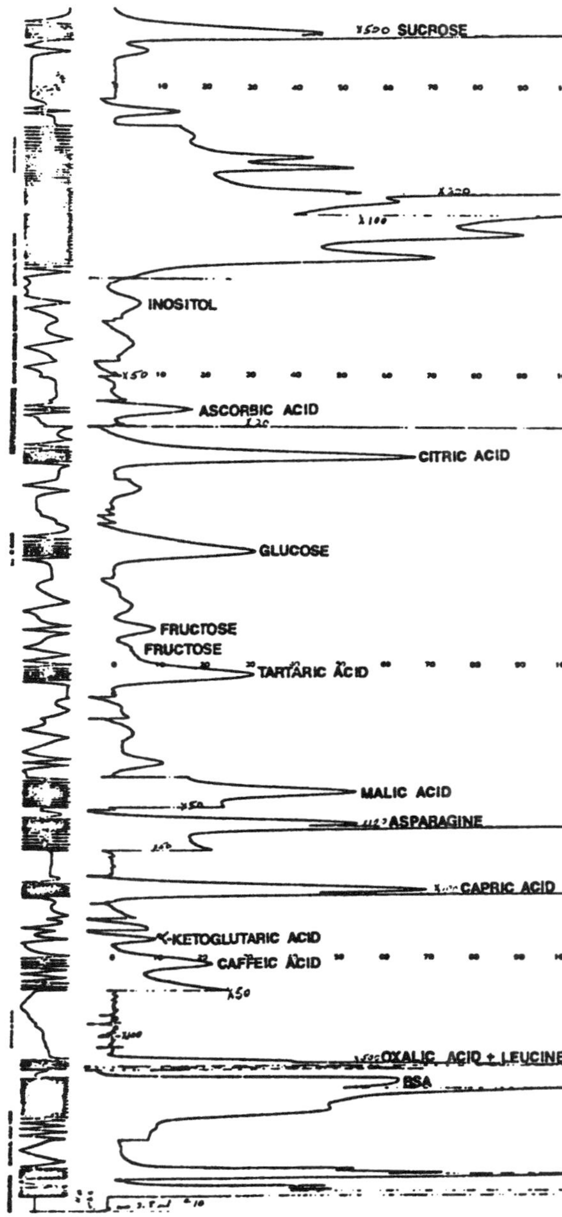

FIGURE 14.1 — Monona Variety

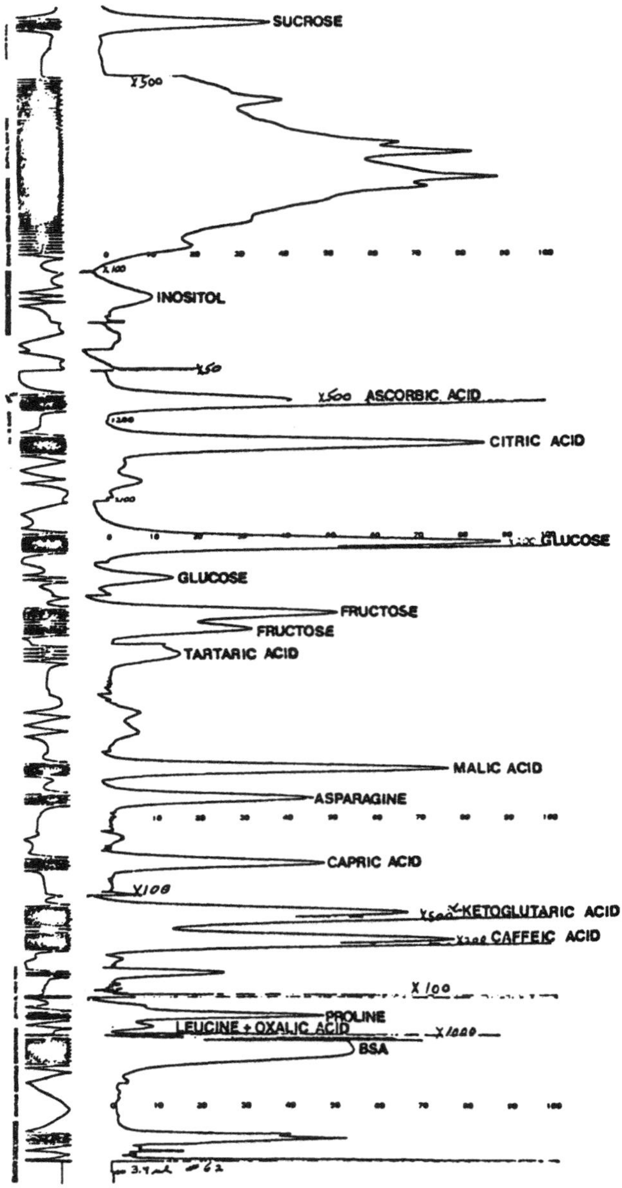

FIGURE 14.2 — Sebago Variety

Man can and should control the amount of respiration in storage by controlling the temperature, humidity and handling practices. When tubers respire, reducing sugars are being broken down to carbon dioxide and water. The breakdown of reducing sugar requires oxygen, therefore, the management of tuber storage areas must control oxygen levels along with temperature to prevent reducing sugar breakdown.

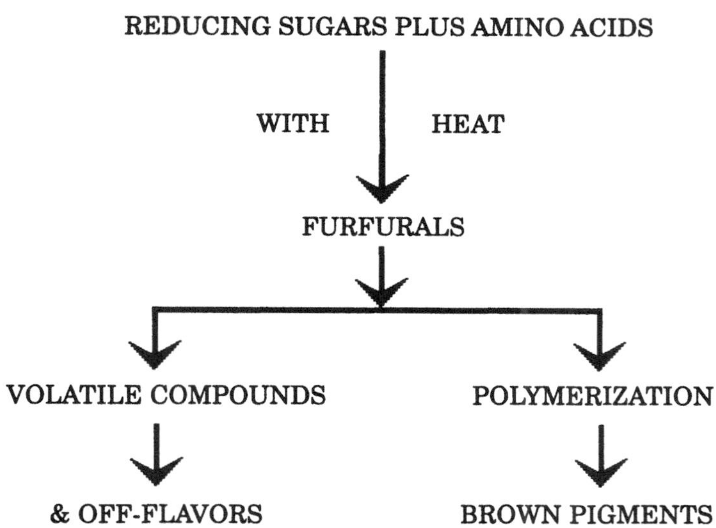

FIGURE 14.3 — Maillard Reaction

The role of the carbohydrates in color formation of finished potato products (dried, chipped and Frozen fries) is important to the manufacturer for producing uniform quality products. Reducing sugars, in particular, acting with amino acids (Malliard Reaction) will produce dark colored chips, fries and dried products very easily. The control of reducing sugars by maturity of the tuber, by storage conditions, and/or by reconditioning periods could eliminate some of the color problems.

Other chemical components of the tuber are very significant to texture and flavor of the finished products. The type of starch has direct bearing on the texture producing fluffy potatoes or waxy potatoes. Again these are varietal differences.

The acid contents are significant in producing different flavors. Chlorogenic acid and caffeic acids differ widely by varieties and they can have an effect on storage as well as flavors of the finished products.

CHAPTER 15

Color and Color Measurement

The color of the processed potato product(s) is almost wholly related to the raw material and the processing practices.

Color is affected by the variety of the tubers in use, how the tubers were grown, fertilized and their maturity at harvest, stress related conditions during growing, harvest and handling of the tubers in and out of storage, and by the processing method that is, time, temperature etc. used to make the finished product. Without question, color is the most difficult attribute of quality to control and maintain in an acceptable range for most processed potato products.

The following are some suggestions for better color control in processed potato products:

1. Know the variety/cultivar of potato that you are using and use only those varieties/cultivars that have indications that they are acceptable for the type of product you are manufacturing.
2. Know the amounts and types of fertilizer that was used on the crop, particularly the Nitrogen levels and be certain that they do not exceed what is required to produce a good crop.
3. At harvest time determine the sugar levels and make certain that the reducing sugars are below 0.15% and, if the tubers are to be stored, that the sucrose level is below 1.5%.
4. Also, at harvest time make certain that the pulp temperatures of the tubers is at or above 50°F (10°C).
5. If in storage, do not move or handle tubers until the temperature is at 70°F (21°C).

6. In test runs for chips and the color is off, that is, above 3 on the Snack Food Association (SFA) 3 point scale: (a) slice the tubers thin, that is, 0.050 to .055"; (b) wash the slices in hot water 170°F (77°C) for 30 seconds before frying; (c) fry at lower temperatures (not to exceed 350°F (177°C) at Inlet, and never let Outlet temper-ature exceed 330°F (165.5°C) as most of the color is developed at the end of the cook.
7. With dried potatoes, all of the above is appropriate except for frying the chips. Here it is important that the temperature of the potato product in the dryer does not exceed the critical temperature. The critical temperature, that is, when the potato temperatures in the drier approach the dryer temperature, must be determined for each style of dried product. The diffusion period in drying is the most critical stage and one must make certain that the product has not case hardened and that the moisture loss is continuous throughout the drying cycle.
8. With French fries, color may be stimulated by adding sucrose to the blanched slices to give them color. One could control this color by careful control of the 1 through 5 suggestions above.

FIGURE 15.1 — Agtron reading 65 and higher.*

FIGURE 15.2 — Agtron reading 55-64.*

FIGURE 15.3 — Agtron reading 45-54.*

Included on the inside, back cover is a color, fold-out, Snack Food Association, ***Potato Chip Standard Color Reference Chart*, of the three photos from pages 126 & 127. This color reference is included with this title for the readers on-the-line, plant use.*

Color for potato chips represents 30 points on a 100 point score. The Snack Food Association has developed a subjective 3 point color chart.

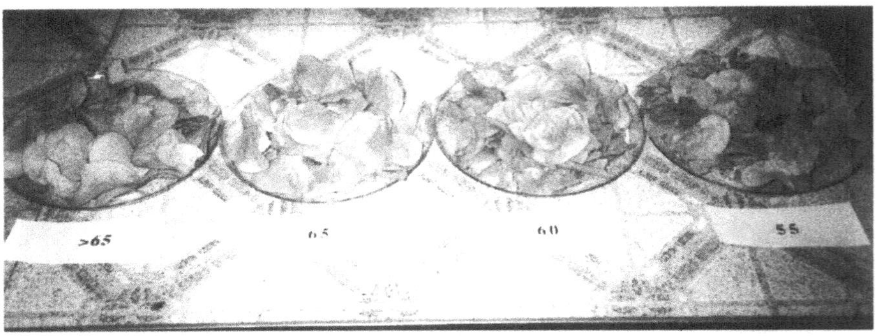

FIGURE 15.4 — Color Variations in Chips

FIGURE 15.5 — Color Score for the Quality Levels of Potato Chips

SFA Color	1	2	3
Color Score	26-30	21-25	11-20
Agtron 90/100	>65	55-64	45-54

This subjective system is a good quick easy benchmark system to use. A 1 on the SFA Color system or a score of 26-30 indicates a light golden color chip while an SFA reading of 4 or a score 10 or less indicates a dark chip and many would say that it is unacceptable. This color chart system should be used with a daylight lamp, preferably a North light, and the observer should not be color blind.

Many prefer an objective benchmark system and the Agtron has met that need. The Agtron is a widely used objective instrument that measures the relative spectral reflectance characteristics of non-homogenous and particulate foods. The capability of the Agtron to provide accurate and reproducible measurements is based on concentric diffuse illumination of the

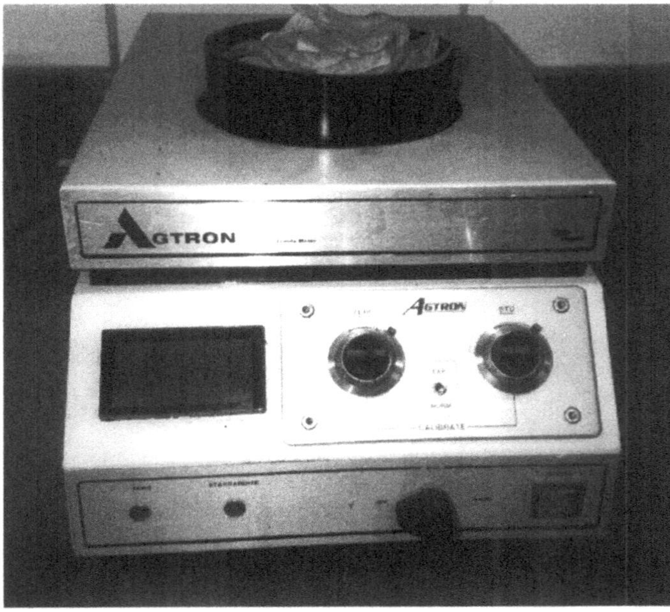

FIGURE 15.6 — Agtron E10 System

sample of chips. The reflected light from the sample represents an average color character of the sample which is largely unaffected by particle size, particle geometry, irregularities, voids, and/or shadows.

There have been several models of the Agtrons on the market. The Model 30 reads the color of the sample as a ratio of green to visible red light (546nm/640nm).

The E5F model is a direct reading reflectance colorimeter which has gained much popularity within the industry. The design provides the ratio of reflectance of a product in two special modes, green and infrared (546nm/811nm). With the E5F, the sample is viewed from above and displays the standard, calibrated at the 90 reading when the sample drawer is out or open.

FIGURE 15.7 — Agtron M30 System

The Model E 10 was designed to replace the Model E5F as it is a microprocessor based instrument and fully automatic self calibrating colorimeter. The instrument displays the color of the chips by rationing the green signal (546 nm) by an infrared signal (811 nm).

In using the Agtron, the cup should be level full, that is, 2 inches deep of chips. They should not be crushed and they need not be covered if the chips fill the cup to eliminate any voids. The readings as shown in Figure 15.6 correspond with SFA subjective color and scores.

The Agtron should be standardized using the 00 "black" disk at 0 and with the 90 "white" disk at 90. Standardization is most important to true readings. Readings are only as accurate as the standardization of the instrument and the proper filling of the cup without crushing the chips.

CHAPTER 16

Texture and Moisture Content

Texture and Factors Affecting

Texture or mealiness (dry flaky texture versus wet or soggy condition) of most processed potato products is a difficult characteristic because so many factors enter into its control and measurement. Some of those factors include the potato, the methods of preservation and the procedures used to measure texture differences.

The potato is probably the number one factor due to the wide differences in total solids and/or moisture content. As previously indicated potato solids are made up of carbohydrates, proteins and fats, although the latter two are in small amounts. The carbohydrates are primarily starch, but may include sugars, pectins and some cellulose. The pectins vary depending in great part on the variety/cultivar being used.

The potato varies greatly in total solids content as previously indicated. High total solids mean less water and, therefore, a more dryer potato. Differences among potato varieties and areas of production go as high as 12 to 15% and this is very important when one is frying or drying the potato.

Therefore, the first characteristic to measure or have facts on is the total solids content of the tuber. Total solids can be measured by drying the sample, that is, measuring the water content. Water and total solids should always equal 100. Normally total solids is measured by measuring the specific gravity and extrapolating the total solids from previously presented tables. The higher the total solids content the better the texture. It is possible, however, to have too high a total solids content, that is, above 1.115 because the potato chips or dried potatoes may become very woody with a difficult bite.

Texture is a factor in potato chips and dried potatoes and it can be affected by the moisture content of the chip and the oil content of the chip. The moisture content of the potato chip should be 1.50% + or – 0.25%. Chips lower than 1.25% moisture content are costly to manufacture and lower yields are obtained. In addition, low moisture content chips will not give bulk in the bag, that is, the bag will not appear to be full. Higher moisture content will shorten shelf life. This same statement is important to dried products. Excess oil, particularly high temperature setting oil will give the appearance and feel of poor texture to the chips.

With frozen French fries, texture is affected by the specific gravity of the tuber, by thickness of cut, but more importantly by processing practices. The use of 2 stage blanching and 2 step frying probably is the greatest factor(s) in control of the texture of the fry. The new additives like starch, sugars and pectins really affect the texture and this may be the future way to control the texture of fries. Most people believe the texture of the fries should

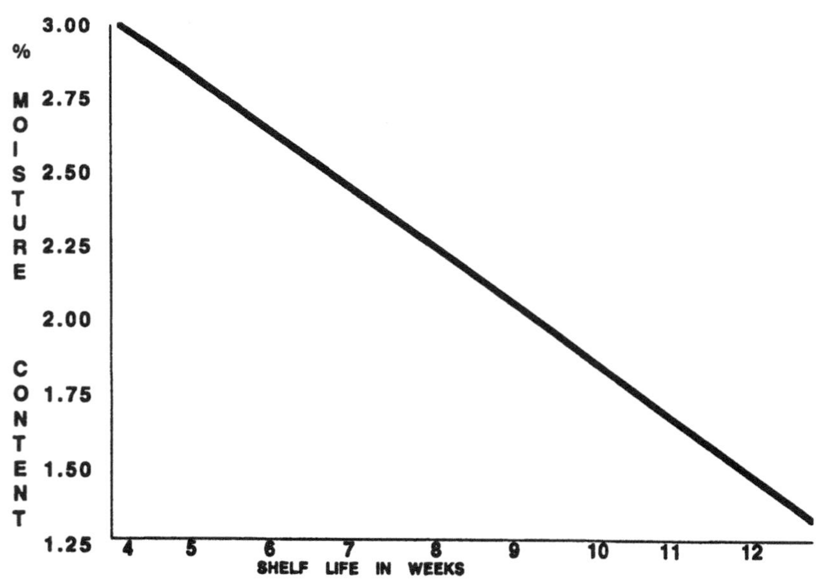

FIGURE 16.1 — Effect of Moisture Content on Shelf Life of Snacks

be crisp and show no noticeable separation from the inner portion of the fry. They should not be limp or show signs of limpness. Further, as with chips the fries should not be oily and the interior portions should not be soggy.

With dried potato products, the texture should be fluffy upon rehydration and free of lumps or case hardened pieces. Dried potatoes can be made from high specific gravity tubers as long as the tuber is not woody in texture. The rehydrated dried potatoes should not be waxy or have a glossy appearance and they should be fluffy when rehydrated.

Canned potatoes are made from low gravity potatoes and the potatoes (whole, sliced, diced etc.) should not slough after cooking. That is, the dice, slice, or whole potato should retain the raw surface and be free of loose starch or washed away surfaces.

Moisture Content and Methods of Measuring

The moisture content of the finished potato chip and/or the dried product is without doubt the most important characteristic about the product as it is an index of processing efficiency. It also, gives information about the thoroughness of the process and the ultimate shelf life of the product. Low moisture content products are more crisp and have a fresher flavor than high moisture content products.

There are many quick methods for determining the moisture content such as the CSC Scientific procedures using a Cenco Balance. However, the Official Moisture content should be determined with the aid of a Vacuum Oven at 70°F (21°C) for 6 hours (see Figure 16.2) or in a Convection Oven at 103°F (39.5°C) for 16 hours. In either case a representative sample is ground to pass thru a 25 mesh screen. Two duplicate 5 gram samples are weighed and placed in drying dishes and accurately weighed (nearest second decimal place) and placed in the oven(s) for the specified times. The convection oven sample should be weighed after the 16 hour drying period and replaced back in the oven until no further loss takes place. The vacuum oven sample is weighed and the moisture content is calculated as follows:

$$\frac{\text{Weight of Sample} - \text{Weight of Dried Sample}}{\text{Weight of Sample}} \times 100 = \%\ \text{Moisture}$$

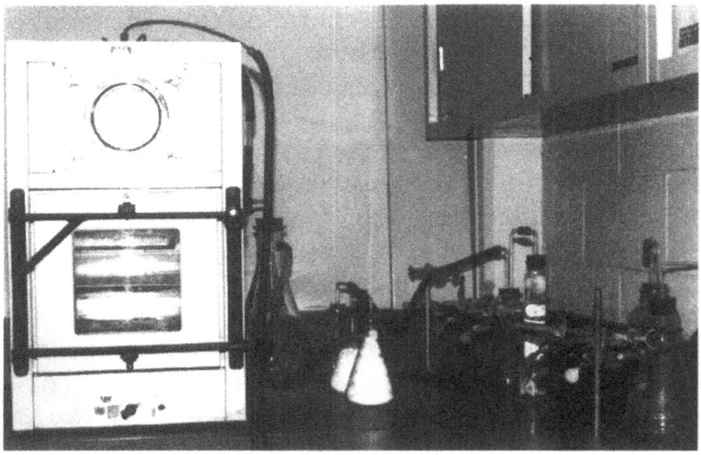

FIGURE 16.2 — Convection Oven

The same calculation is used for the Convection sample after it is thoroughly dried.

Quick methods are used by many and they are OK once they have been standardized with the oven methods. Quick methods will give moisture results in a matter of a few minutes and they are as accurate as the sample is. Mettler, Ohaus and CSC Scientific are good examples. The CSC Scientific Moisture instrument is my preference for a stationary standardizing instrument as it is easy to use and results can be read direct from the instrument. Generally these instruments are top loading using infra red heating. Care must be exercised in to standardize the heat time, size of bulb (Ohaus) and height of lamp(s)

Many production lines in a factory need on-line readings of the moisture content using instantaneous techniques to measure and control moisture and oil readings as the product is being manufactured. These continuous measurements allow the operator to better control the production process by assuring uniform consistent qualities on each line. Some of these on-line instruments make their measurements with near infrared light. These are reliable and repeatable techniques and they do not destroy or contact the product on-line. These instruments can be set-up to read moisture, oil content and amounts of seasonings being applied to the product.

FIGURE 16.3 — In-Line Moisture Measurement

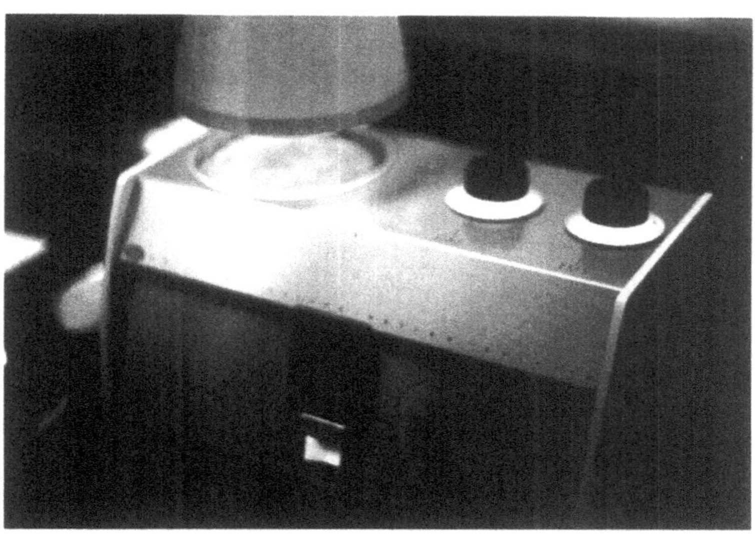

FIGURE 16.4 — Quick Moisture Instrument (Courtesy Ohaus)

FIGURE 16.5 — CSC Digital Moisture Balance

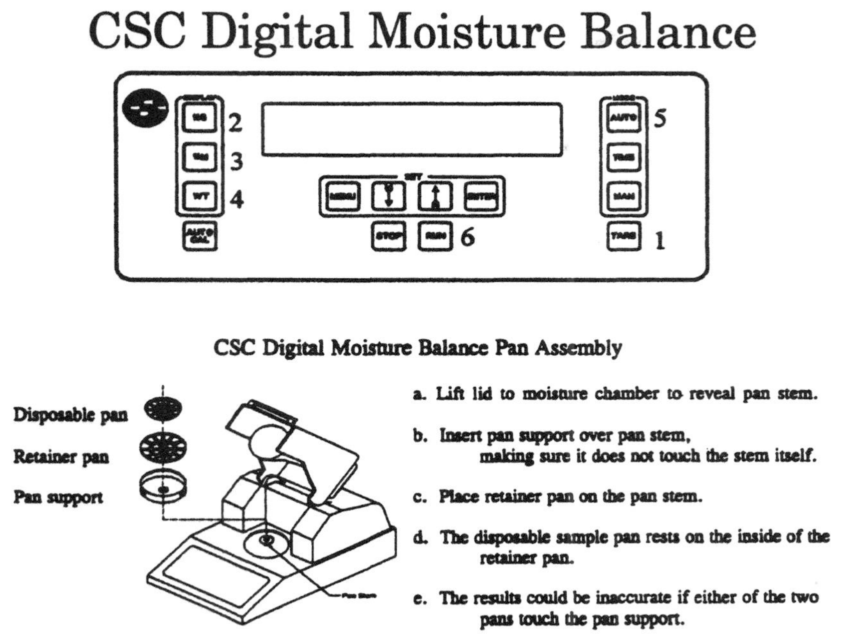

FIGURE 16.6 — Digital Moisture Balance Reading Instruction (Partial)

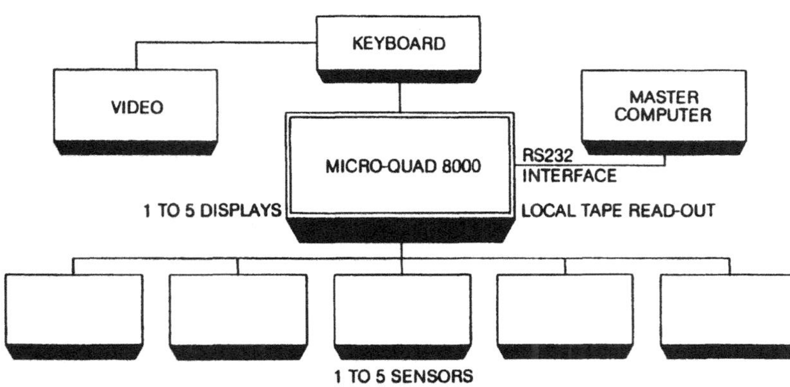

FIGURE 16.7 — Constant Process Monitoring

CHAPTER 17

Defects in Potato Products

Defects are not wanted in potatoes or potato products. They cost time and money to remove.

They take away from the good appearance of the finished products.

They are difficult to remove and cause losses in product yield due to peel loss, trim loss, sort loss, and off-colored products.

They are unsightly and they turn some people off from using the different products.

Defects need to be eliminated for the good and growth of the potato processing industry.

Many of the bacterial, fungus, and virus diseases found in and on potatoes are seed transmitted. The major concern are the wilts and rots. Tubers that are infected can go over a grading table unnoticed, but when the tuber is peeled they become very noticeable and a serious problem to remove with great losses. Potatoes that look normal going into storage's may come out with dry rot because tuber infection occurs during storage unless care is taken to remove the infected tubers.

Most of the injury which cause physical defects in potatoes and processed potato products occurs during the harvesting and handling of potatoes from field to storage and out of storage. The bruised or defective areas shows up as a darkened area on the finished product unless removed during the preparation of the tubers for further processing. Chemical methods for detection of surface bruises are available and should be used to pinpoint damage (see below). Black spot and other discolored areas may be observed after peeling. Internal black spot, pressure bruises, harvest cracks and shatter bruises are examples. A blackspot bruise usually does not penetrate more than a quarter of an inch while shatter bruise appears as a fissure or a series of fissures with discoloration at the edge and it may penetrate deeper. These types of defects are a financial loss to the producer as a lower price may be offered for the load. Further, there is an increased weight loss in the load.

Physical defects can be minimized by controlling the following five factors:

1. Proper soil moisture at the time of harvest, that is, between 60 and 80% of the field capacity.
2. Delaying harvest for as long as 20 days after vine kill.
3. Harvesting only when the soil temperature is above 50°F (10°C).
4. Proper operation of the harvester as to down the row speed, keeping the chains at full capacity, and elimination of excessive drops.
5. Operating the loader, unloaded, and storage operations to prevent drops and roll back.

Understanding this phase of the potato business can control bruises and eliminate these types of defects.

Another group of defects are the physiological defects. Examples are blackheart, growth cracks, heat necrosis, and internal sprouting. Generally they are caused by improper storage temperatures, lack of ventilation and aeration, and inadequate moisture or relative humidity.

Pathological defects are usually caused by bacteria, fungi, or viruses. Examples are bacterial soft rot, scab, dry rot, blight and necrosis.

Entomological defects are caused by insects: aphids, leafhopper, fleabeetles, potato tuber worms and wireworms.

All diseases and insect problems should be controllable by proper cultural practices including the correct usage of pesticides. Any evidence of diseased or insect damaged tubers, if not removed during preparation for processing will show up in the finished products. They are unsightly, unwanted and can cause the product to be rejected at the point of purchase or consumption.

Recently, the supply industry has developed some very good

FIGURE 17.1 — In-Line Color Sorter For Removal Of Defective Product

and highly efficient in-line sorters to remove defects. Installation of this equipment can clean up the defect problem and present products that are free or absence of defects. However, the removal of these defects costs time, money, and energy. The removal of defects with the in-line sorters are as efficient as the operator wants them to be, but as one operator has stated, "I can't afford to waste all that product".

Bruise Detection on the Raw Potato

Chemicals and Equipment

Catechol (1 % solution)
Knife
Potato Peeler

Procedure

1. Select a representative sample (minimum of 10 tubers)
2. Wash the tubers
3. Immerse the tubers in the catechol solution for 1 minute

4. Remove the tubers and allow them to dry
5. Determine the number and extent, if any, of the damaged tubers by noting the dark red or purplish areas on the surface of the tubers
6. Evaluate the severity of the bruise as follows:
   a. Skinning — one stroke of knife removes all visible damage
   b. Slight bruise — two strokes of the knife removes all visible damage
   c. Serious bruise — more than two strokes are required to remove all damage.

Since the catechol bruise detection method only identifies bruises that break the skin, such a method does not reflect total damage or other types of defects. Internal bruising (blackspot) can only be detected after peeling.

Testing for Freedom of Defects in Potato Chips

Defects in potato chips have reference to pieces of peel, black discoloration on portions of the chips, green discoloration, internal discoloration of the chip, or other harmless extraneous material. Defects are determined by examining the potatoes and classifying those that are minor (less than 1/4 inch) and those that are major (greater than 1/4 inch). The actual chips that are defective are counted, but more than one defect on a slice is counted only as one defective slice. Blisters are considered a defect and they, too, are counted the same way as reported as percent of chips that are blistered. Blisters are believed to be cultivar related and may be concentrated in specific growing areas, probably due to environmental conditions. Hash is due, in part, to handling chips and are considered defects. The pieces of hash are counted in the same way, that is, each piece is considered as one chip and the percentage calculated as above.

1. Obtain a 2 qt. representative sample;
2. Count the total number of chips and separate any chips which show defects, blisters and hash as follows:

Defect Classification	Minor	Major
Group I		
Discolored appearance which adversely affects the chip to a noticeable degree, i.e., 1/4 inch or less.	x	
Discolored appearance which adversely affects the chip to a degree which is objectionable, i.e., more than 1/4 inch in area		x
Group II		
Blemished area including peel, internal discoloration, or harmless extraneous material which adversely affects the chip, i.e., 1/4 inch or less.	x	
Blemished area including peel, internal discoloration, or harmless extraneous material which seriously affects the chip, i.e. more than 1/4 inch in area.		x

FIGURE 17.2 — Defect Evaluation of Potato Chips

		Maximum Percent of Defects			
Class	Score	Minor	Major	Blister	Hash
1-A	18-20	0-5	0-3	0	1
2-B	16-17	6-10	4-5	1	3
3-C	11-15	11-15	6-8	2	5
4-D	6-10	16-20	9-12	3	7
5-F	0-5	> 20	> 12	> 4	>10

3. Calculate the percentage of chips for each class and record results as shown in Figure 17.2.

   Note: If a chip has more than one defect, count it as one defect, but if it is a major defect count it as a major defect.

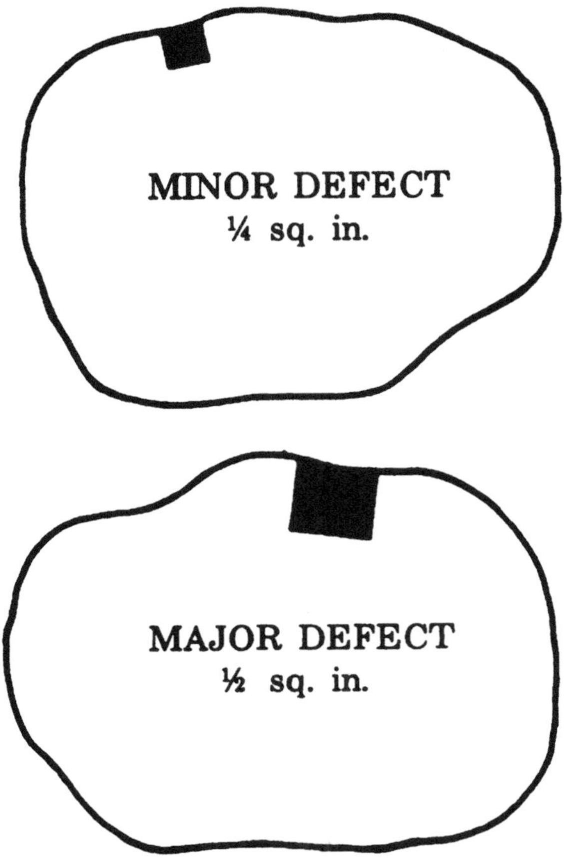

FIGURE 17.3 — Minor Defect/Major Defect

A blister is any chip with a blister greater than a 20 mm circle. Blisters are counted and reported as percentage per sample.

Hash is any piece of chip that is smaller than 1 square inch . It is separated from the sample and counted and the percentage calculated by dividing the number of hash chips by the total number of chips in the sample and multiplying by 100 to obtain % hash.

CHAPTER 18

Flavor of Processed Potato Products

The potato is a very bland product, and it's what we do to it in processing that makes it appealing and entices one to buy it again. In the case of canned whole potatoes, its the salt or brine that we cook them in. In the case of canned potato salad, its the onions, vinegar, salad dressing and spices that we use. In the case of potato chips and frozen French fries, it is the cooking oil that is used and in some cases the seasoning additives that may be used.

The potato may have its own flavor from the effects of chemicals in production, processing and packaging that may affect the flavor of the finished product. These are usually unwanted flavors and they can be controlled by man in the production, processing and storage of the product. One such chemical that has caused serious problems has been the inheritance problem of glycoalkaloids. Varieties/cultivars with glycoalkaloids above 20 ppm are not to be used because of their effect on off-flavors.

Reports of off-flavors from spoilage in handling potatoes continue to crop up and loads must be rejected if one finds soft rot in loads as the flavor is unbearable in the finished products. Other rots may occur in potato piles creating unwanted flavors.

Chemicals to control specific insects and/or diseases may create problems in the cooked product, whether canned, dried or fried. Compounds with benzene in their make-up have created serious off-flavor problems after processing.

Modern technology continues to add flavors to the potato to make the finished product more relevant and interesting to the ultimate user. These flavor changes and/or additions are not to hide things, but to make them more appealing and in some cases make new products for the ultimate satisfaction of the user.

Oil Effect

Oils are used to hasten the cooking or drying of the potato. Each oil has its own flavor and many manufacturers have developed their market around the flavor of their specific oil. Peanut oil is an excellent flavored oil and markets in central Ohio have been developed around peanut oil. Corn oil is a big favorite in the Chicago area, cottonseed oil is big flavor in the South while Soya oil is used in the midwest, etc. The industry, also, uses blends of oil to get wider marketing coverage due to their marketing practice and area of distribution of their products. The new Olean oil, although it has little flavor, is being heavily promoted and may find a future niche not because of flavor but because of no recoverable oil or calorie increase by the consumer that eats this product.

Regardless, oils are most important to the manufacturer and there is a large trend to better control of the oil content. In the early days of the chip industry, percent oils in the finished product were much higher than they are today. During World War II the Quartermaster Corp had a tolerance for oil content at 46% on potato chips. Today, no one should really make chips with that high an oil level. The industry is pushing for oils at 30% or lower for meeting nutrition suggestions and really for better flavor and consumer satisfaction.

If oils are not fresh and not properly turned over or used up in manufacture, they may develop some off-flavors over time to the chips and fries. Turnover of the oil in efficient operations should be once during every shift, however, turnover should be based on any build up of free fatty acids in the oil. The free fatty acid may increase in the cooking oil if it is not turned, if it is not properly filtered, if air is mixed with the oil as in cavitating pumps, if in contact with metals like copper or brass or with other oxidizing metals, or if it is abused and left exposed to air for periods of time. Free fatty acid build up is a sure sign of oil breakdown.

Good manufacturers do protect their oil when it is not being used by emptying the cookers and holding the oil in vertical tanks after filtering it and covering the filtered oil with a nitrogen blanket to protect it from the air. The care of the oil is without doubt the most important factor when operating a chip or fry plant as the oil will impart flavor to the finished product. The flavor imparted may be the difference between acceptance or rejection of the finished product.

Free Fatty Acid Measurement

The amount of free fatty acids (ffa) in the frying media indicates the degree of breakdown of the frying oil. Shelf life, flavor, and quality of finished chips and fries can be directly related to the amount of free fatty acids. It is mandatory that fryer operators understand oil breakdown and to know when it has reached an unusable ffa value.

Fresh oil should have a free fatty acid value of less than 0.05% while used oil can go as high as 0.5% and be quite satisfactory, but when the free fatty acid value approaches 5.0% one should know that this will impart off-flavors and the resultant product can be quite unacceptable. Therefore, one should know how to measure free fatty acids. The Snack Food Association has a kit including the chemicals and apparatus needed to measure the ffa.

The kit contains an automatic burette assembly, complete with a 10 ml capacity burette, glass reservoir bottle (for 0.1 N sodium hydroxide solution), 2-hole rubber stopper, rubber bulb, glass burette tip, pinch clamp and rubber tubing for connecting tip to burette. Also, included in the kit is a 250 ml wide mouth Erlenmeyer flask, 50 ml graduate cylinder with a mark at 32.5 ml for measuring the oil sample which is to be used, and a dropping bottle for the alkaline blue 6B indicator solution (saturated solution of alkali 6B powder in isopropyl alcohol).

Procedure to measure the ffa content

1. Fill the graduate to the 50 ml mark with isopropyl alcohol and empty into the Erlenmeyer flask.
2. Add 20 drops or more of the alkali blue indicator to the Erlenmeyer flask to obtain a dark blue color with the alcohol.

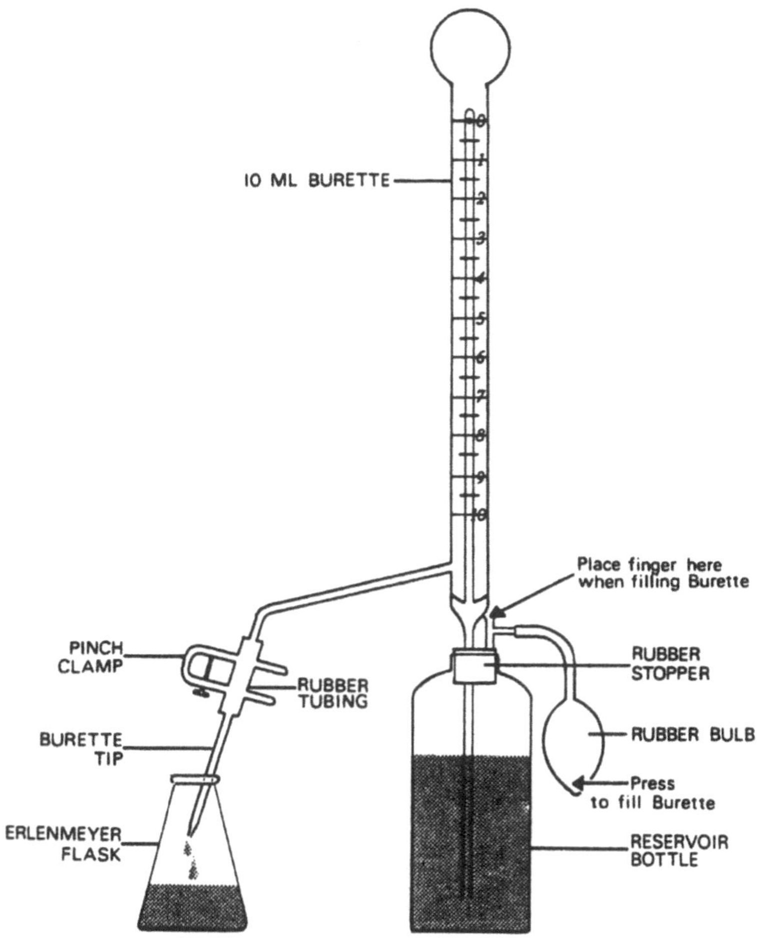

FIGURE 18.1 — Free Fatty Acid Testing Aparatus
(Courtesy Snack Food Association)

3. While shaking the flask add 0.1 N sodium hydroxide solution from the burette, one drop at a time, until the color of the isopropyl alcohol first changes from blue to red.
4. Fill the graduate to the 32.5 ml mark with oil (if not hot oil, heat the oil to 150°F (65.6°C) to be tested for the ffa content. Empty the oil completely into the red colored isopropyl alcohol in the Erlenmeyer flask. Shake the flask vigorously to mix the oil and alcohol. Color will change to blue.
5. Fill the burette to the top mark with 0.1 N sodium hydroxide.
6. Titrate the oil sample in the Erlenmeyer flask with sodium hydroxide solution from the burette by opening the pinch clamp slightly and shaking the Erlenmeyer flask vigorously until the color changes from blue to red. Continue to shake the flask and add the sodium hydroxide solution until the color does not disappear.
7. Read the burette to the nearest 0.1 ml to determine the amount of sodium hydroxide used in the titration. Each whole number (1.0) is equal to 0.1% ffa, each small division (0.1 ml) is equal to 0.01% ffa. For example, 5.2 ml equals 0.52% ffa.

The ffa test is used to evaluate the quality of the oil as received, the quality of the oil as it is being used, and the quality of the chips for shelf life. For the latter, the oil must be expressed from the chips and filtered before titrating.

Percent Oil Content

Another important test is the amount of oil absorbed by the product. This test is important because oil is expensive and, secondly, oily products are not desirable. They are considered greasy by the consumer.

Oil on chips and frozen French fries may be extracted from the finished product by leaching the oil from the products using given solvents. The use of fat soluble solvents may be very hazardous, therefore, much care must be exercised by the analyst and they must use great caution and a hood for the extraction. The most common procedure in use is called the Bailey-Walker process.

The Bailey-Walker Test

The Bailey-Walker process requires a good analytical balance, mortar and pestle to grind the sample, graduate cylinder for measuring the solvent, fat soluble solvent, electric air oven to dry the sample after extraction, condenser vials, paper thimble to hold the sample, and an electric hot plate with water circulating system.

FIGURE 18.2 — Oil Extraction Press

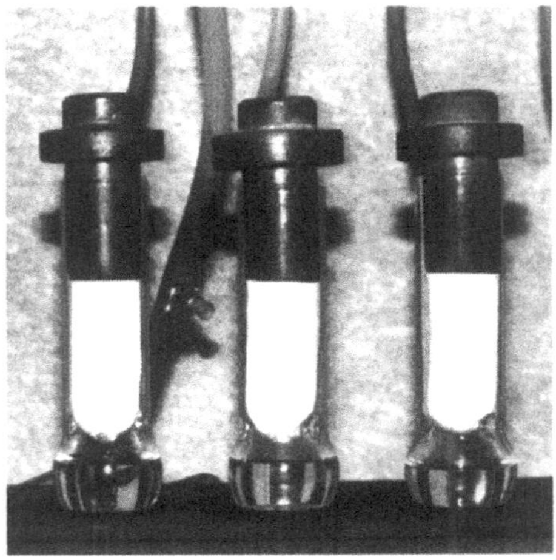

FIGURE 18.3 — Baily-Walker Oil Extraction Collection Vials

Procedure:

1. Weigh the oven-dried thimble.
2. Using the mortar and pestle, grind the sample.
3. Weigh out a 10 gram sample and place in the preweighed thimble.
4. Add approximately 40 ml of solvent, such as, petroleum ether to the condenser vial and then place the thimble with the ten gram sample in it into the vial. Use enough ether to fill up the vial just below the top of the thimble.
5. Turn hood on and start the water circulating through the condenser.
6. Turn the hot plate on high until the sample begins to boil, then turn the hotplate to a low temperature and reflux for 1 hour.
7. At the end of 1 hour, remove the thimble and dry for 30 minutes in the circulating air oven at 212°F (100°C).
8. Weigh the thimble and sample and calculate the percent

oil by dividing the weight of oil by weight of the sample times 100 and report as % oil.

Oils can be expressed from the sample using a Carver Press as follows:

1. Weigh out 10 grams of representative sample into sample chamber.
2. Place the entire assembly in the press and pump it up at a rate of about 1 stroke every 2 seconds until 15,000 lbs. of pressure is reached.
3. Allow 20 seconds for the pressure to drop down and then pump it back up to 15,000 lbs. pressure.
4. Set the timer for exactly 3 minutes.
5. At the end of 3 minutes, release the pressure and remove the sample cake from the bottom of the chamber using care to obtain all the cake portion (do not include any of the oil portion) and weigh the sample cake only.
6. The weight of the oil less cake times 100 equals the percent oil. The percent oil pressed from the sample will be around 50% of the true amount of the adsorbed oil. For each oil used, one must correlate the expressed oil with the actual extracted oil, therefore, it is essential to use the extraction procedure to obtain the true oil value.

   Note: This is a quick procedure and one must exercise care in always following the details of the procedures and always having the sample at the same temperature. It is a safe procedure and does not require an expensive hood, use of solvents, etc.

Salt and Seasonings

Flavor of chips are affected by the addition of salt and seasonings. Salt has been a flavor enhancer since Biblical times as referred to in the Book of Job and Gospel According to Matthew. The Snack Food Association recommends that if salt is added that it be added at the rate of 1.75% plus or minus 0.25% for potato chips. There are several methods to evaluate the salt content, but the most common method is to use the DiChromat.

Salt Determination

1. A 25 gram representative sample is weighed and added to 250 ml of distilled water and blended to make a slurry.
2. The slurry is filtered to obtain 120 ml of filtrate.
3. The filtrate is poured thru a previously standardized DiChromat and the reading is obtained on the digital readout.

This is a very simple method and is as accurate as the sample. Variations in salt content do occur and it is up to the operator to maintain the salter always in good working condition.

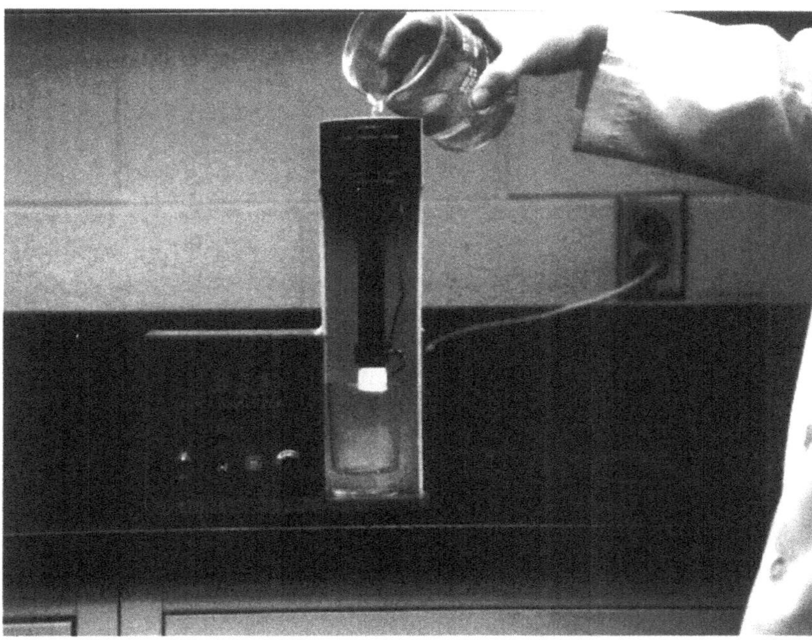

FIGURE 18.4 — DiChromat Salt Determination

Seasonings may be added to potato chips to create new and enjoyable and different flavors to potato products. At the present time, about 1/4th of the snacks are seasoned. The major seasoning is BBQ and this varies from very mild to very hot. Following

BBQ in terms of percent usage are Vinegar and Onion, and Cheese.

Seasonings may be applied by tumbling the seasoning with the chips, by sifting the seasoning on top of the product, by blowing the seasoning on, or by making a slurry of the seasoning and spraying this on the product. Of course, many seasonings are added to potato chips, French fries and other potato products by the ultimate user, either as dips, toppings or mixed with the product.

Every manufacturer has their own concept of what a seasoner is made up of and many seasoning use carriers such as salt or whey. Therefore, the user needs to obtain from the purveyor of the seasoning his or her method of analysis to detect the amount of seasoning applied or to apply to the product. Seasonings can be removed from the chip by dissolving the seasoning in water or other solvents and then quantified using a spectrophotometer. Each seasoning from different suppliers may have different wavelengths for detection. Therefore, the supplier or manufacturer of the seasoning should provide the method of extraction and the method of quantifying.

Generally, seasonings are applied at 6 to 8%, however, this will vary depending on the carrier in the seasoning, the manufacture of the seasoning, and the actual seasonings. With modern methods of control, the levels should be consistent for any given label.

Flavor Evaluation

Flavor evaluation is one of the more difficult tasks to be performed on snacks or any potato product because each individual has his or her own preferences and differences. Flavor should be evaluated on the basis of acceptance or no acceptance, but because of the wide range of flavor additives to many potato products, it is usually scored. Scoring of flavor should be done with a panel of judges knowledgeable in sensory work. The combined judgment of several individuals will give a more accurate story of the flavor and may save a firm a whole lot of time and money when adding flavor additives to their products.

Triangle Evaluation

All tasting should be done in an odor-free room and each panel member should work independently and each should record their decision separately. If a new product is being developed, the triangle test should be used wherein the panel member is asked to identify the odd sample from the duplicate sample. After they have identified their odd sample Vs the duplicate sample, they should then express their reason for the odd Vs the duplicate by scoring the two samples. Obviously if they cannot correctly identify the odd samples from the duplicate samples there is no difference. If they correctly identify the odd sample from the duplicate samples and score one over the other, the samples do differ and action can be taken whether or not to proceed further in the development and makerting of the product. Triangle testing of products is an excellent method of separating the vital one from the trivial many products.

Scoring for Flavor

In the manufacturing of potato products, the order of each day is to manufacture some more like yesterday, that is, assuming yesterdays run was satisfactory. So, on a regular declared time of day, a tasting session is conducted. For the actual tasting, the production manager, the sales manager, the QA manager and, at least, 7 others should make up a panel to individually review the results of the past 24-hour run (assuming a three-shift operation).

The samples should be presented to the panel from each hours operation by coding them using a 3-digit or 3-letter code only, that is, 435, 691, 827, etc. or AJM, KCZ, PWB, etc. The samples should be placed in cups or on trays. The panel members should have a score card, a cup of water to rinse their mouth with, and an empty cup for the mouth rinse. The panel members should be asked to score the sample on a 10 point scale or some other numerical scale that is used for that product. In practice, I like to use a duplicate sample in the series to be scored, obviously, with a different code. If the panel members cannot find the duplicate sample or do not score it close to each other, there is little or no difference in the samples or that panel member does not have the ability to distinguish the differences. After each session the results should be tabulated. Samples scoring above or below the average should be noted for their differences and a determination from QC data should confirm the reason for these differences.

Some people like to calculate the Least Significant Difference after interpreting the results using Analysis of Variance or the 'F' test. This is no doubt a good practice for some kinds of data, but for production runs it is not needed. I do, however, like to interpret data by keeping individual records on each panel member over time and, of course, charting the days run by making an X bar and R chart. These charts are valuable over time to see what the firm is doing in terms of the quality of the products they are making.

Flavor and in some cases odors (off-odors) can be most helpful in describing the quality of manufactured products. Off-flavors or off-odors are not desirable and should not be permitted in the market place. Acceptable flavors and odors are necessary to maintain and increase the status of that product in the market place. The customer expects certain flavors from your products and his/her expectations must be met all the time for repeat sales.

CHAPTER 19

CGMP and Food Plant Sanitation

The Current Good Manufacturing Practices (CGMP) sets the background for food plant sanitation. The details are published as part 110 of the Code of Federal Regulations. There are several parts of the regulations that are important to potato processors and to modern day food plant sanitation.

The Current Good Manufacturing Regulation

Part 110.5 states that the criteria and definitions in this part shall apply in determining whether a food is adulterated (1) within the meaning of section 402(a)(3) of the Federal Food, Drug and Cosmetic Act in that the food has been manufactured under such conditions that it is unfit for food; or (2) within the meaning of section 402(a)(4) of the act in that the food has been prepared, packed, or held under insanitary conditions whereby it may have become contaminated with filth, or whereby it may have been injurious to health.

Part 110.10 states that the plant management shall take all reasonable measures and precautions to ensure the following:

(a) Disease control. Any person who, by medical examination or supervisory observation, is shown to have, or appears to have, an illness, open lesion, including boils, sores, or infested wounds, or any other abnormal source of microbial contamination by which there is a reasonable possibility of food, food-contact surfaces, or food-packaging materials becoming contaminated, shall be excluded from any operations which may be expected to result in such contamination until the condition is corrected. Personnel shall be instructed to report such health conditions to their supervisors.

(b) Cleanliness. All persons working in direct contact with food, food contact surfaces, and food-packaging materials shall conform to hygienic practices while on duty to the extent necessary to protect against contamination of food. The methods for maintaining cleanliness include, but are not limited to:

(1) Wearing outer garments suitable to the operation in a manner that protects against contamination of food, food contact surfaces, or food packaging materials.

(2) Maintaining adequate personnel cleanliness.

(3) Washing hands thoroughly (and sanitizing if necessary to protect against contamination with undesirable microorganisms) in an adequate hand-washing facility before starting work, after each absence from the work station, and at any other time when the hands may have become soiled or contaminated.

(4) Removing all insecure jewelry and other objects that might fall into food, equipment, or containers, and removing hand jewelry that cannot be adequately sanitized during periods in which food is manipulated by hand. If such hand jewelry cannot be removed, it may be covered by material which can be maintained in an intact, clean,and sanitary condition and which effectively protects against contamination by these objects of the food, food-contact surfaces, or food-packaging materials.

(5) Maintaining gloves, if they are used in food handling, in an intact, clean and sanitary condition. The gloves should be of an impermeable material.

(6) Wearing, where appropriate, in an effective manner, hair nets, headbands, caps, beard covers, or other effective hair restraints.

(7) Storing clothing or other personal belongings in areas other than where food is exposed or where equipment or utensils are washed.

(8) Confining the following to areas other than where food may be exposed or where equipment or utensils are washed: eating food, chewing gum, drinking beverages, or using tobacco.

(9) Taking any other necessary precaution to protect against contamination of food, food-contact surfaces, or food-packaging materials with microorganisms or foreign substances including, but not limited to, perspiration, hair, cosmetics, tobacco, chemicals, and medicines applied to the skin.

(c) Education and training. Personnel responsible for identifying sanitation failures or food contamination should have a background of education or experience, or a combination thereof, to provide a level of competency necessary for production of clean and safe food. Food handlers and supervisors should receive appropriate training in proper food handling techniques and food-protection principles and should be informed of the danger of poor personal hygiene and insanitary practices.

(d) Supervision. Responsibility for assuring compliance by all personnel with all requirements of this part shall be clearly assigned to competent supervisory personnel.

Part 110.20

(a) Grounds. The grounds about a food plant under the control of the operator shall be kept in a condition that will protect against the contamination of food. The methods for adequate maintenance of grounds include, but are not limited to:

(1) Properly storing equipment, removing litter and waste, and cutting weeds or grass within the immediate vicinity of the plant buildings or structures that may constitute an attractive, breeding place, or harborage.

(2) Maintaining roads, yards, and parking lots so that they do not constitute a source of contamination in areas where food is exposed.

(3) Adequately draining areas that may contribute contamination to food by seepage, foot-borne filth, or providing a breeding place for pests.

(4) Operating systems for waste treatment and disposal in an adequate manner so that they do not constitute a source of contamination in areas where food is exposed.

If the plant grounds are bordered by grounds not under the operator's control and not maintained in the manner described in paragraph (a)(1) through (3) of this section, care shall be exercised in the plant by inspection, extermination, or other means to exclude pests, dirt, and filth that may be source of food contamination.

(b) Plant construction and design. Plant buildings and structures shall be suitable in size, construction and design to facilitate maintenance and sanitary operations for food-manufacturing purposes. The plant and facilities shall:

(1) Provide sufficient space for such placement of equipment and storage of materials as is necessary for the maintenance of sanitary operations and the production of safe food.

(2) Permit the taking of proper precautions to reduce the potential for contamination of food, food-contact surfaces, or food-packaging materials with micro-organisms, chemicals, filth, or other extraneous material. The potential for contamination may be reduced by adequate food safety controls and operating practices or effective design, including the separation of operations in which contamination is likely to occur, by one or more of the following means: location, time, partition, air flow, enclosed systems, or other effective means.

(3) Pertains to fermented foods.

(4) Be constructed in such a manner that floors, walls, and ceilings may be adequately cleaned and kept clean and kept in good repair; that drip or condensate from fixtures, ducts and pipes does not contaminate food, food-contact surfaces, or food-packaging materials; and that aisles or working spaces are provided between equipment and walls and are adequately unobstructed and of adequate width to permit employees to perform their duties and to protect against contaminating food or food-contact surfaces with clothing or personal contact.

(5) Provide adequate lighting in hand-washing areas, dressing and locker rooms, and toilet rooms and in all areas where food is examined, processed, or stored and where equipment or utensils are cleaned; and provide safety-type light bulbs, fixtures, skylights, or other glass suspended over exposed food in any step of preparation or otherwise protect against food contamination in case of glass breakage.

(6) Provide adequate ventilation or control equipment to minimize odors and vapors (including steam and noxious fumes) in areas where they may contaminate food; and locate and operate fans and other air-blowing equipment in a manner that minimizes the potential for contaminating food, food packaging materials, and food contact surfaces.

(7) Provide, where necessary, adequate screening or other protection against pests.

Part 110.35 deals with sanitary operation including cleaning and sanitizing,

110.37 deals with sanitary facilities and controls,

110.40 is concerned with equipment and utensils,

110.80 deals with processes and controls, and

while 110.93 deals with warehousing and distribution stipulating that food shall be under conditions that will protect food against physical, chemical, and microbial contamination as well as against deterioration of the food and the container.

It should be noted that the complete regulation is available from FDA and is published in my book, *CGMP/Food Plant Sanitation* (available from CTI Publications, Inc.).

Likewise, an interpretation of the regulation and food plant sanitation principles are covered in the above book, therefore, the reader is referred to this book for further details on this subject including sanitation practices, cleaning, pests, and how to plan a sanitation program.

To be successful and to keep a plant in good sanitary condition, it is most important that the designated sanitarian have a committee (team) to work with. This committee should consist of a representative of top management, the production supervisor, the human resource person, quality assurance director, and if unionized, a union representative. These people should meet regularly and tour the facility while the plant is operating and they should make recommendations where and if needed. The sanitation director should be the official recorder and he or she should participate fully in the tour and discussion. It is essential that records be taken and maintained to assure the status of the facility and when and where changes must be made. The Sanitation Overview Report is completed by the team while the Evaluation Report Figure 19.2 is completed by the Sanitarian on a daily basis. Please note I have set this up for a chip plant and used a 1000 point scale to evaluate the operation. This form should be completed and filed daily with the committee.

When and if the inspector comes from the municipality, the state or from the federal level the sanitation director should accompany the food inspector throughout his or her tour of the facility. If the food inspector takes samples, the firm representative should also take samples. If the food inspector is allowed to take pictures, the firm representative should do like wise. At the conclusion of the official inspection and before exiting the premises he or she will give the firm representative an "I observed" written form. This should be accepted and discussed to clarify any points made in the report. Next it behooves the sanitation committee to meet and clarify and cleanup any areas of concern. It is far wiser to stay ahead of any problems rather than try and fight them.

The following paragraph is a personal interpretation of things

I would be concerned about when the inspector comes: What does the outside of the plant, parking lot, etc., look like? Having been an inspector and visiting many food plants and noting someone emptying his/her car ash tray in the neatly marked parking lot leaves the wrong impression. Secondly, seeing a large flag pole, but someone forgot about putting the flag up or upon entering the front door and seeing the bait boxes at the front door turned over. These above impressions before I enter the facility leave one with great skepticism about the food operation. Further, lack of trimmed shrubbery, signs for the cat or dog to read, or doors propped open are all suspicious acts that leave the wrong impression. Upon entering the plant and visiting the rest room and seeing the trash overflowing from the trash container, toilet seats covered with toilet paper, wash basins dirty, etc., leave another bad impression of the facility. Thirdly, after waiting for several minutes before being let into the plant and upon entry noting moldy ceilings, trash on the floor, employees standing around, etc., leaves another wrong impression of the facility. Remember, we have not seen the food operation yet and that first impression is most important. It does not take much effort to make the first impression a good impression before seeing the food operation itself.

FIGURE 19.1 — Food Plant Survey & Sanitation Inspection Report

DATE:__________ PLANT:____________________ RECORDER:____________

PREMISES:

- OUTSIDE AREAS
- PARKING LOT
- RECEIVING AREA
- SHIPPING AREA
- WASTE DISPOSAL
- OTHER

BUILDINGS:

- OFFICE
- FACTORY
- STORAGE
- WAREHOUSE
- OTHER

MATERIALS:

- POTATOES
  - VARIETY
  - SHIPPING AREA
  - VEHICLES
  - HANDLING SYSTEM
  - SAMPLING SYSTEM
  - GRADING PRACTICE
  - SPECIFIC GRAVITY
  - SIZE
  - DEFECTS, IF ANY
  - CONDITION
- OIL
  - DELIVERY SYSTEM
  - TYPE OF OIL
  - FFA ON ARRIVAL
  - HOLDING TIME ON PREMISE
  - NITROGEN
  - CONDITION
- PACKAGES
  - CONDITION
  - TEMPERATURE OF STORAGE ROOM
  - RH OF STORAGE ROOM
  - CONDITION

FIGURE 19.1 — Food Plant Survey & Sanitation Inspection Report - Continued

- OTHER INGREDIENTS
  - HOLDING TIME
  - TEMPERATURE
  - RH
  - CONDITION

<u>PREPARATION FOR PROCESSING:</u>

- HOLDING AREA
- DESTONING
- PEELING
- WASHING
- SORTING
- TRIMMING
- SLICING
- WASHING
- BLANCHING
- FRYING
- INSPECTING
- SALTING
- SEASONING
- SORTING/PICKING
- HOLDING
- PACKAGING
- WEIGHTS
- CLOSURES
- CASING

<u>WAREHOUSING:</u>

- CLEANLINESS
- METAL DETECTION
- OTHER

<u>PERSONNEL:</u>

- CLEANLINESS
- CLOTHING
- HAND WASHING
- JEWELRY
- HAIR RESTRAINTS

<u>CONTROLS AND AUDITS:</u>

- PRODUCT EVALUATIONS
- LABORATORY SAFETY & PRACTICES

FIGURE 19.2 — WAGCO's CGMP's Evaluation Form

FIRM				DATE
AREAS	UNIT OPERATIONS	POINTS	SCORE	EXPLANATION
OUTSIDE	Entrance	10		
	Perimeter/landscape	10		
	Building(s)	10		
	Receiving yard	10		
	Scales	10		
	Flume & conveyors	20		
	Loading/Unloading docks	10		
	Parking lot(s)	10		
	Waste management	10		
		100		
STORAGE(S)	Floors	10		
	Walls	10		
	Ceilings	10		
	Temperature	10		
	Relative Humidity	10		
	Ventilation	10		
	Product handling	20		
	Pests	10		
		90		
PREPARATION	Flumes/Conveyors	10		
	Elevators/Belts	10		
	Washers	10		
	Peelers/Corers/Slicers	10		
	Walls	10		
	Floors	10		
	Ceilings	10		
	Ingredient Protection(Metal, etc.)	20		
	Equipment Maintenance	30		
	Sanitary Practices	30		
	Pests	10		
		160		

FIGURE 19.2 — WAGCO's CGMP's Evaluation Form - Continued

PERSONNEL	Cleanliness	30		
	Head, facial coverings	20		
	Smoking, food, drink, etc.	10		
		60		
QUALITY ASSURANCE	Sampled upon receipt	10		
	Sampled in process/formulation	10		
	Sampled after processing	10		
	Sampled in warehouse before shipment	10		
	Codes/coding	10		
	Laboratory records up-to-data	10		
	Handling out-of-specification products	10		
	Written recall program up-to-date	10		
	Written Laboratory Manual/Procedures	10		
	QA data book up-to-date	10		
	Retained sample storage area in use	10		
	Sanitation Inspection/Report records	10		
		120		
MANAGEMENT	Mgtm by walking around (MBWA)	10		
	Evidence of vision	10		
	Promoting published Values	10		
	Flowcharts with Critical Control Pts.	10		
	Supplier guarantees/AQL's	10		
	Personnel training programs	10		
	Customer Expectation/Complaint records	10		
		70		
OTHER	Other			
TOTAL SCORES		1000		
EXPLANATION WHERE APPROPRIATE, THAT IS, SCORE NOT EQUALING POINTS POSSIBLE				

CHAPTER 20

Using HACCP to Help Assure the Safety of Potato Products

The food industry generally has one goal and that is repeat business. Repeat business is built on the fact that the industry produces, prepares, processes and packages safe wholesome food. The customer expects this and that is why repeat business builds manufacturing firms. The manufacturer does this in well defined programs using continuous surveillance with up-dated teams practicing modern technology and good communications.

Responsibilities for food safety start with management as food safety is truly a management function. The authority for carrying out in-plant safety functions must be assigned to responsible persons.

Food safety requires good records covering all facets from procurement through preparation, processing details, and packaging including the human factor for each function. The records must be accurate and they must reflect the actual operating conditions at the time specified. The records must be in ink and the process operators must sign or initial as the records are developed. Further, all records must be on appropriate forms and they must be approved by the respective supervisor. The records must prove that the product is safe and wholesome.

The FDA and the USDA working in concert with industry through their respective associations, such as NFPA and SFA, have developed a NEW system known as Hazard Analysis Critical Control Point or HACCP. This system has been defined as the identification of sensitive ingredients, critical control process points and control limits, and relevant human factors as they affect food product safety. The technique maximizes the beneficial impacts of management decisions by first identifying potential products risks. Those risks identified as critical are then evaluated and strategically located control points are established to monitor those product and process risks to assure the safety of the food. Thus, HACCP is preventive in nature and it protects the consumers from exposure to potential food hazards.

Penalties for failure in food safety are most severe: they include both fines and imprisonment.

- HACCP is a management tool for assuring food safety. It should be the top priority of any food firm.
- HACCP requires that everyone in management work together to identify all potential hazards and risks in the operation.
- HACCP requires the development of appropriate prevention programs to control the process and apply all corrective actions where needed.
- HACCP demands training of all personnel to understand the identification and monitoring requirements. Management and all employees must train together to design and control the process through identification of potential hazards and know how to prevent the system from going wrong. A strong HACCP training program followed by empowerment of manufacturing employees will provide the motivation for success.
- HACCP requires a strong commitment by management with a thorough communication up and down the lines of authority including all employees.
- HACCP requires the right product design and process control plan(s) using a multidiscipline team (personnel from procurement, thru production, engineering, quality assurance, sanitation, and management) that truly embrace the concept.

Employees must want to take ownership of the process for HACCP to be beneficial and successful.

HACCP has five pre-HACCP steps, as follows:

1. Bringing together all the firms HACCP resources to write the program and establish the technology to be used including the flow of the product through the various unit operations in the process used in the plant. This includes a

team consisting of the Plant Engineer, Production Manager, Sanitation Director, Quality Assurance Manager, Food Microbiologist and others deemed knowledgable about the products and processes. This team must understand the formulation of the products, its method of procurement, processing, packaging, and distrbution. Further, the team must understand what kind(s) of abuse could take place along the line including those from the consumer.

2. A complete description of the product and its method of distribution for each product processed.
3. A complete list of the ingredients and all raw materials and how they are purchased, stored and held prior to processing.
4. A process flow diagram to identify all the steps used to prepare a product from storage through final shipment. These steps should help to identify unit operations in the establishment where hazards could occur. All steps in the process should be identified to make sure that the steps listed describe "what really occurs in producing and processing the product". In other words, "Walk the Plant".
5. Meet the regulatory requirements of Sanitation Standard Operating Procedures (SSOP). These are based on the Federal Food Code and generally follow the CGMP's as previously described.

These preliminary steps are then followed by HACCP's clearly defined seven steps as follows:

1. Identification of all the HAZARDS associated with the product and the process. These include the biological, chemical, and physical hazards that could be associated with the product or the process including CFR 21 Part 113 "Thermally Processed Low Acid Foods" for the canned potato products.
2. Identification of the Critical Control Points (CCP's) in the process. A CCP is a point, step, or procedure at which control can be applied and a food safety hazard can be prevented, eliminated, or reduced to acceptable levels.

3. Establishment of the CRITICAL LIMITS (CL) for preventive measures associated with each identified CCP's. When a CCP is identified, a CL is then established to signify whether a CCP is in or out of control.
4. Establishment of procedures to monitor the CCP's.
5. Establishment of the Corrective Action to be taken when monitoring shows that a CL has been exceeded.
6. Establishment of effective record keeping systems that document the HACCP system.
7. Establishment of procedures to verify that the HACCP system is working.

Following are five forms that could be helpful in establishing HACCP and making Corrective Actions when and if needed (see Forms 1, 2, 3, 4, and 5 (Corrective Action).

- HACCP should be implemented now in every food plant by developing a plan that emulates the process with clearly marked critical control points followed by computerizing your data to monitor and provide corrective action instructions to focus on potential hazards during the process to assure that safe food is being produced.
- HACCP is cost effective and it reduces the need for testing of finished products to assure that the products being manufactured are safe.
- HACCP can be most helpful and beneficial in today's processing and marketing plan for any food firm.
- HACCP is a system of prevention Vs inspection. It depends on people and their understanding of the written program. Further, the people must understand the system and properly execute the plan for HACCP to be successful. Lastly, the people must understand the system and verify that the product was made safely and is safe for consumption.
- HACCP is pro-active and not re-active. People are the key to making HACCP work right. For example, for every possible hazard that may be encountered, there needs to be a specific control put into place to prevent the problem. People can and must do this.

- HACCP is really nothing new in terms of procuring our product and ingredients; preparing, processing, and packaging the product correctly, and proper labeling of the product during the marketing of your products right. It really just reemphsizes what we should have been doing all along in producing safe and wholesome foods.
- HACCP is a system to assure that the product was made safely and records must be available to prove that it was made safe.
- HACCP is a system which is becoming widely accepted as the discipline of choice for augmenting CGMP's and traditional quality assurance methods to ensure safety of processed foods.

FIGURE 20.1 — Basic Concerns Relative To Potential Hazards

Item	Concern	Method of Control
Growers	Pesticides Herbicides Defoliants	{Always use on *right* crop, {at *right* time, and in {*right* amount
Haulers	Sanitation and Time/Temperature	PreWash and Clean > 50°F or 10°C with time as short as possible
Suppliers	Ingredients	Comply with specifications as to Chemical, Microbiolgical and Physical requirements
Preparation	Stones, Sticks, Metals, HEM*	Utilize eliminators, i.e., Sieves, Screens, Magnets, Traps, and/or Detectors
Processing	Time and Temperature	Automatic Controllers, Failure Alarms
Containers	Integrity of Closures	Visual and Tear Down Inspections

*Harmless Extraneous Materials

FIGURE 20.2 — HACCP Process Steps

FORM 1

HACCP PROCESS STEPS / HAZARDS / PREVENTIVE MEASURES / CCP'S

Process Steps	Potential Hazards	Preventive Measures	CCP Description

FIGURE 20.3 — HACCP Monitorinmg Actions

FORM 2

HACCP MONITORING / CORRECTIVE ACTIONS

Process Step/CCP	Critical Limits	Monitoring Procedures					Corrective Actions
		What	How	When	Where	Who	

FIGURE 20.4 — HACCP Recordkeeping

FORM 3

HACCP RECORD KEEPING / VERIFICATION

Process Step/CCP	Records	Verification Procedure

FIGURE 20.5 — HACCP Master Record Sheet

FORM 4

HACCP MASTER RECORD SHEET BY PRODUCT

Process	CCP	Hazard	Critical Limit	Monitor	When	Who	Corrective Action	Documented	Verification

FIGURE 20.6 — CCP Corrective Action

FORM 5

CRITICAL CONTROL POINT / CORRECTIVE ACTION

Plant: ______________________________ Date: ____________

CCP No.: ______________________________ Rev. Date: ____________

Unit Operation: ______________________________

Critical Control Point: ______________________________

Operation Task: ______________________________

____

____

Responsibility: ______________________________

Reason for Control: ______________________________

____

Method of Control: ______________________________

____

Operating Procedure: ______________________________

____

____

Corrective Action: ______________________________

____

Review Procedure: ______________________________

____

Developed by: ______________________________ Date: ____________

____________	____________	____________	____________
Product Mgr.	*Production Mgr.*	*QA Mgr.*	*Plant Mgr.*

CHAPTER 21

Potato and Potato Chip Terms and Terminology

BRUISE — Common defect on potatoes, particularly mechanically harvested and/or stored potatoes. Generally differentiated from other mechanical defects in that the tuber skin has not been broken.

COLD CHIPPERS — A term used to define potato cultivars which have the genetic potential to produce light colored chips directly out of long term storage (6 to 8 months) at 40°F (4.4°C).

CULL — A rejected product because of inferior quality.

CULTIVAR — A term used to identify a variety or new line of a given commodity. Originally used to identify numbered lines, but now includes named varieties.

DEFECT — Anything that is not wanted, an imperfection, or something that is foreign to the product.

ENZYME — A complex mostly protein product found in all living cells that induces or speeds chemical reactions without itself being permanently altered. Enzymes must be controlled to prevent browning or oxidation of potatoes or rancidity of chips.

FABRICATED CHIPS — A style of chips wherein potato flour and/or other flours are made into a masa, extruded into a given shape, and fried or dried.

FAT — An edible mixture of triglycerides that are solid at ambient temperatures.

FIRE POINT — The temperature at which a fat or oil will burn when heated above its given temperature.

FLASH POINT — The temperature at which a fat or oil will flash when a flame is passed over the surface of the oil. This is generally some 20 to 30 degrees below the Fire Point.

FREE FATTY ACID — The free acids formed when a fat or oil begins to break down.

GUMMING — Formation of a gum or sticky material resulting from continued heating of fats or oils. It is produced by oxidation and polymerization of the fat or oil and represents breakdown products which collect on heated surfaces.

HYDROMETER — A floating device developed by Ora Smith and SFA to determine the specific gravity or density of potatoes.

HYDROGENATION — Hydrogenation is a process used to change a liquid oil into a semisolid or solid fat at ambient temperatures. Hydrogenation provides stability to fats, it results in higher melting points and longer shelf like without rancidity.

KETTLE CHIPS — Sliced potatoes that may or may not be washed and fried at a somewhat lower temperature than regular chips and usually are more curled due to a system of frying and may have more oil adsorbed due to longer fry time at lower temperature.

LOT — Any number of containers or loads of the same size and type which contain a product of the same type and style located in the same warehouse or conveyance, or which, under in-plant inspection or in-process results from which consecutive production within a plant, and which is available for inspection at any one time.

LYE — A strong alkaline solution, usually Caustic Soda (sodium hydroxide) used for cleaning fryers to remove polymerized fats, etc.

MAILLARD REACTION — A dark color to chips resulting from the chemical interactions between sugars and proteins.

MAJOR DEFECT — A defect, that is, discolored area, that is greater than $1/2$" in size.

MATURE/MATURATION — The phenomena of ripening in plants. Potatoes are considered mature when the vine is deceased and the tubers are free from feathering. Chemically it is when the sucrose content of the enlarged tuber is at its lowest point, hopefully, less than 1.00%.

MINOR DEFECT — A defect on a potato chip that is less than 1/4 inch in size.

MOISTURE — In raw potatoes, it is referred to as water content, but in chips it is referred to as moisture content and it should be below 3%, preferably < 2% for good shelf life.

MONOSACCHARIDE — A 6-carbon sugar, such as, glucose.

OIL — An edible mixture of triglycerides that are liquid under ambient temperature.

OXIDATION — A chemical reaction involving the addition or combination of oxygen with other chemicals.

OXYGEN SCAVENGER — An antioxidant. Sometimes used in oils coming from the refiner or it may be added to salt and seasonings or packaging materials to help prevent oxidation of the oil or the finished product.

PATHOLOGICAL DEFECTS — Defects caused by micro-organisms. Examples are soft rot, scab, dry rot, blight, and necrosis.

PERIDERM — The outer cells, commonly referred to as peel, of the potato. In the potato these outer cells or peel are dead cells and provide some fiber and minerals to the consumption of same.

PEROXIDE VALUES — Fats consist of saturated and unsaturated acids in combined forms with glycerin. The unsaturated acids are susceptible to oxidation. Oxygen can add to a fatty acid to form peroxides or hydroperoxides. The peroxide value is a measure of the amount of these products.

PHYSIOLOGICAL DEFECTS — Defects caused by handling or storage practices, such as, black heart, growth cracks, heat necrosis, and internal sprouting.

PITH — The central area of the potato tuber or the main storage area.

POLYMERIZATION — The forming of gummy substances when frying due to oil breakdown resulting in greasy or oily finished products.

POTATO CHIPS — Thin slices of potatoes, fried in oil and may be lightly salted. They got their start as "Saratoga Chips", named for their birth place, the resort at Saratoga Springs, NY. In order to satisfy a difficult customer George Crum, the native American Indian Chef of Moon's Lake House, prepared French fried potatoes so thinly they became mere crisps. The term Crisp is still used in England for what we call chips.

RANCIDITY — An oxidative deterioration in foods containing fats whereby a typical off-odor and/or off-flavor is produced.

REDUCING SUGARS — A six carbon sugar like glucose or fructose which is easily oxidized. Reducing sugars react with given amino acids and cause potatoes to fry dark. Reducing sugars are formed from sucrose and should never exceed 1.5% for use in the manufacture of acceptable colored chips.

RELATIVE HUMIDITY — Ratio of water vapor present in the air to the quantity that would be present if the air were saturated at the same temperature. Potatoes should be stored with the RH > than 90%.

SAMPLE — It is a representative part of something selected at random from a lot and used for inspection.

SAMPLING — The act or practice of selecting samples from a lot or load used for inspection or acceptance or rejection.

SAPONIFICATION — The hydrolysis of mono-, di-, or triclycerides with caustic or alkali to form free glycerol and fatty acids in the form of soaps.

SHELF LIFE — The duration of time a product will remain acceptable the user.

SLICE THICKNESS — The thickness of a potato slice, generally measured in thousands of an inch. The normal thickness will range from 0.050" to >0.80". Thin chips, that is, less than 0.055" of an inch will break easier and have more oil adsorbed to the chip for a given weight of chips.

SMOKE POINT — The temperature at which a fat or oil gives off a thin continuous stream of smoke or a sign of imminent breakdown of the fat or oil.

SPECIFIC GRAVITY — A measure of total solids content of a product. Potatoes are composed of solids and water. Potatoes normally range from a total solids content of 13 to 37%, which means the water content ranges from 63 to 87%. Lower water content potatoes or higher solid content potatoes are preferred.

STARCH — A white, odorless, tasteless complex carbohydrate produced in plants as an energy source. The primary component of potatoes is starch.

SUCROSE — A 12-carbon disaccharide sugar found in potatoes and made up of one molecule of glucose and one molecule of fructose. Sucrose is chemically reduced in the potato plant to form starch.

SWEETENING — A process of starch breaking down to sucrose and in turn to glucose and fructose after a given resting period. The objective of storage is to prevent the potatoes, following harvest and storage, from sweetening.

STYLE — The characteristic or form of manufacture of potato chips, such as *regular* or *flat*, that is, plain slicing of the potato versus *wavy* or *ridged* cutting. The latter may have more oil adsorption if cut at the same thickness.

TOCOPHEROL — A naturally occurring antioxidant found in some vegetable oils that retards the onset of rancidity.

TURNOVER — The rate at which fat is used up during a frying operation. It is affected by the amount of fat adsorbed per unit of fried food, the number of units being fried during the given heating period and the ratio of amount of fat to fried food in the fryer at one time.

TUBER — Potatoes are an enlarged underground stem of the potato plant bearing buds or eyes in their axils from which sprouts and/or new plants may be found.

APPENDIX 1

Snack Food Association Recommendations for Vehicle Sanitation

- Vehicles delivering and transporting raw materials to snack food plants shall be clean, in good sanitary condition, and in, compliance with the requirements of the Sanitary Food Transportation Act Of 1990 and implementing regulations of The Department of Transportion (DOT).
- The cleanliness and condition of vehicles are the responsibilities of the transportation firm and/or the owner of the equipment.
- All vehicles shall be thoroughly cleaned prior to loading and free from any chemical, physical, microbiological, and/or odor contamination.
- All vehicles shall be free of any debris prior to loading and during hauling. Debris includes, but is not limited to, pallets and empty containers.
- All vehicles shall be inspected by the person loading the vehicle for foreign materials including, but not limited to, glass; oil deposits; defective floors, ceilings (if enclosed) and side walls; insects and insect signs, such as trails, webbing, or cast skins; and rodents and rodent signs, such as chewing, droppings, and urine stains.
- All vehicles shall haul produce, particularly potatoes, as gently as possible. Rough handling and improper temperatures may cause adverse chemical reactions which may increase sugar content or cause other biological changes.
- All vehicles must carry accurate thermometers for checking the pulp temperatures of potatoes. The temperature of potatoes destined for chipping facilities should always be maintained at or above storage temperature, but should not exceed 75°F.

- If combustion-type heaters are used during winter months, ventilation allowing fresh air exchange must be provided to circulate the heated air as carbon dioxide may build up in vehicles.

- Any vehicle containing human or animal waste shall not be loaded or used to haul any materials to or from a snack food plant.

- Inspection records must be completed on all incoming materials. See attached form for minimum recommended information.

VEHICLE SANITATION INSPECTION RECORD

DATE: ______________

PRODUCT: ____________________________

NAME AND IDENTIFICATION OF CARRIER:

____

____

CLEANLINESS OF OUTSIDE OF CARRIER:

____

____

ARE SEALS ACCEPTABLE? ______________

IF NOT, DESCRIBE: _____________________________________

UPON OPENING DOORS, ARE THERE ANY OFF-ODORS?

YES _____ NO _____

IF YES, DESCRIBE: _____________________________________

TUBER: PULP TEMPERATURE _______ CONDITION ______

CARRIER CONDITION: ACCEPT ______ REJECT ______

REMARKS: ____

____

____

INSPECTED BY: ________________________

APPENDIX 2

United States Standards for Grades of Potatoes for Processing[1]

[*Effective July 10, 1963*]

[Redesignated at 42 FR 32514, June 27, 1977]

Source: 28 FR 5607, June 7, 1963, unless otherwise noted.

---

[1] Compliance with the provisions of these standards shall not excuse failure to comply with the provisions of the Federal Food, Drug and Cosmetic Act, or with applicable State laws and regulations.

DEFINITIONS

2851.3416	Similar varietal characteristics.
2851.3417	Fairly well shaped.
2851.3418	Fairly smooth.
2851.3419	Soft rot or wet breakdown.
2851.3420	Sprouts.
2851.3421	Foreign material.
2851.3422	Damage.
2851.3423	Usable piece.
2851.3424	Serious damage.

AUTHORITY: The provisions of this subpart issued under secs. 203, 205, 60 Stat. 1087, as amended, 1090 as amended; 7 U.S.C. 1622, 1624.

GRADES

§2851.3410 U.S. No. 1 Potatoes for Processing.

"U.S. No. 1 Potatoes for Processing" consists of potatoes of one variety or similar varietal characteristics which are fairly well shaped and fairly smooth, which are not frozen, and are free from freezing injury, blackheart, late blight, southern bacterial wilt, ring rot, soft rot, wet breakdown, sprouts, foreign material, and free from damage by any other cause, and which meet a size specification as provided in §2851.3413. Unless otherwise specified the potatoes have a specific gravity of not less than 1.070 and a glucose content of not more than one-tenth of 1 percent or fry to a medium brown[2] or lighter color as determined in accordance with the procedures set forth in §2851.3414. (See §2851.3415.)

§2851.3411 U.S. No. 2 Potatoes for Processing.

"U.S. No. 2 Potatoes for Processing" consists of potatoes or usable pieces of potatoes of one variety or similar varietal characteristics which are not frozen, and are free from freezing injury, blackheart, late blight, southern bacterial wilt, ring rot, soft rot, wet breakdown, sprouts, foreign material, and free from serious damage by any other

---

[2] The color referred to is illustrated by plate 10.0 YR 7/8 in the Munsell book of Color. Individual matte finish plates of this and lighter colors may be purchased from the Greytag-Macbeth Co. , 617 Little Britain Road, New Windsor, NY 12553.

cause, and which meet a size specification as provided in §2851.3413. Unless otherwise specified the potatoes or usable pieces of potatoes have a specific gravity of not less than 1.070 and a glucose content of not more than one-tenth of 1 percent or fry to a medium brown[2] or lighter color as determined in accordance with the procedures set forth in §2851.3414. (See §2851.3415.)

UNUSABLE MATERIAL

§2851.3412 Unusable Material.

"Unusable material" consists of the defective portions of potatoes in excess of that permitted in U.S. No. 2 Potatoes for Processing, and potatoes which are frozen, affected by freezing injury, soft rot, wet breakdown, blackheart, late blight, southern bacterial wilt, or ring rot, or which are seriously damaged by internal sprouts, internal discoloration, hollow heart or which are otherwise defective and cannot be trimmed in preparation for processing to form usable pieces.

SIZE

§2851.3413 Size.

(a) The minimum size, maximum size, or range in size shall be specified in connection with the grade in terms of diameter or weight of the potato or usable piece, or in accordance with one of the classifications given in the following table: *Provided*, That if no size is specified the diameter shall be not less than $1^1/_2$ inches. (See §2851.3415.)

(b) Diameter means the size designation in terms of inches or eighths of an inch indicating the greatest crosswise dimension at right angles to the longitudinal axis of a potato or usable piece that will pass lengthwise through a round opening. A potato having a designation of 2 inches is one which will not pass through a round opening 2 inches in diameter but will pass through a round opening $2^1/_8$ inches in diameter.

(c) Weight means the size designation in terms of whole ounces indicating the weight of the potato or usable piece. A potato having a designation of 10 ounces is one which weights at least 10 ounces but less than 11 ounces.

TABLE I

Size Classification	Round or Intermediate Varieties*		Long Varieties**	
	Min.	Max.	Min.	Max.
Extra Large	4 inches	....	16 ounces	
Large	3 inches	4 inches	10 ounces	16 ounces
Medium	$2^1/_4$ inches	$3^1/_4$ inches	4 ounces	10 ounces
Size B	$1^1/_2$ inches	$2^1/_4$ inches	$1^1/_2$ inches	$2^1/_4$ inches
Size C	1 inch	$1^1/_2$ inches	1 inch	$1^1/_2$ inches
Creamers	$^3/_4$ inches	$1^1/_2$ inches	$^3/_4$ inches	$1^1/_2$ inches
Size A	$1^7/_8$ inches minimum with 60% or more $2^1/_4$ inches and larger		$1^7/_8$ inches minimum with 40% or more 6 ounces or larger	

*Irish Cobbler, Katahdin, Sebago, Kennebec, Pontiac, Red McClure, Pungo, or other similar varieties are considered round or intermediate shaped varieties.

**Russet Burbank, Early Gem, White Rose, or other similar varieties are considered long varieties.

TEST PROCEDURES

§2851.3414 Test Procedures.

Tests to determine specific gravity, glucose content and fried color shall be made in accordance with the procedures set forth in this section. The potatoes used shall be taken at random from a composite sample drawn from containers throughout a load, or a comparable sample from a bulk load or storage bin.

a) *Specific Gravity.* Specific gravity shall be determined by either the hydrometer[3] method, or by calculation from the weights of the sample in air and in water, with equipment which has been tested and calibrated to give accurate results. The reading obtained from each specific gravity test shall be corrected for temperature

[3] A hydrometer designed specifically for determining the specific gravity of potatoes, and the Chip Color Tester, a tape specifically designed for making glucose tests, are available from Potato Chip Institute, International, 946 Hanna Building, Cleveland, OH, 44115.

variations as prescribed by Table II of this section. The specific gravity for any lot of potatoes shall be the average of at least 3 such corrected readings on separate tests from the composite sample.

(1) *Temperature correction.* The pulp temperature of the potatoes and the temperature of the water shall be recorded immediately before testing and the specific gravity reading corrected as indicated in the following table:

TABLE II — Correction Factors For Specific Gravityof Potatoes*

(Corrected to Zero Base of 50° Tuber Temperature and 50° Water Temperature)
[Water Temperature Degrees Fahrenheit]

Tuber Temp.	38	40	45	50	55	60	65	70	75	80
38	-.0021	-.0020	-.0018	-.0018	-.0020	-.0023	-.0029	-.0038	-.0047	-.0056
40	-.0017	-.0016	-.0014	-.0014	-.0016	-.0019	-.0025	-.0034	-.0043	-.0052
45	-.0009	-.0008	-.0006	-.0006	-.0008	-.0011	-.0017	-.0026	-.0035	-.0041
50	-.0003	-.0002	0000	0000	-.0002	-.0005	-.0011	-.0020	-.0029	-.0038
55	+.0001	+.0002	+.0004	+.0004	+.0002	-.0001	-.0007-	.0016	-.0025	-.0034
60	+.0004	+.0005	+.0007	+.0005	+.0002	-.0002	-.0004	-.0013	-.0025	-.0034
65	+.0005	+.0006	+.0008	+.0008	+.0006	+.0003	-.0003	-.0012	-.0021	-.0030
70	+.0006	+.0007	+.0009	+.0009	+.0007	+.0004	-.0002	-.0011	-.0020	-.0029
75	+.0007	+.0008	+.0010	+.0010	+.0008	+.0005	-.0001	-.0010	-.0019	-.0028
80	+.0008	+.0009	+.0011	+.0011	+.0009	+.0006	0000	-.0009	-.0018	-.0027
85	+.0009	+.0010	+.0012	+.0012	+.0010	+.0007	+.0001	-.0008	-0017	-.0026
90	+.0010	+.0011	+.0013	+.0013	+.0011	+.0008	+.0002	-.0007	-.0016	-.0025
95	+.0011	+.0012	+.0014	+.0014	+.0012	+.0009	+.0003	-.0006	-.0015	-.0024
100	+.0012	+.0013	+.0015	+.0015	+.0013	+.0010	+.0004	-.0005	-.0014	-.0023

*To apply correction factor, change, actual specific gravity reading by adding or subtracting the appropriate factor according to the plus or minus sign.

(b) *Glucose Content.* The glucose content is determined by a test based on enzyme reaction as indicated by color changes in a specifically prepared tape[2] applied to the cut surfaces of potatoes in a sample. In determining the glucose content not more than 5 percent of the test tape in contact with the cut tuber may indicate the presence of more than one-tenth of 1 percent glucose.

(c) *Fried Color Determination.* The test to determine compliance with fried color requirements is made by frying slices from sample potatoes in deep fat having a temperature of 350°F at start of frying, and no less than 300°F during the remainder of the frying process. A lot of potatoes meets the specified fried color requirement when not more than 5 percent of the area of any fried slice is distinctly darker than the specified color.

TOLERANCES

§2851.3415 Tolerances.

(a) For the purpose of determining compliance with one of the foregoing grades or of other specifications the following tolerances, by weight, are provided in order to allow for variations incident to proper grading and handling:

(1) Defects:

(i) U.S. No. 1 Potatoes for Processing: 10 percent for potatoes which fail to meet the requirements of this grade but not more than one-half of this tolerance, or 5 percent, for unusable material, including therein not more than 1 percent for potatoes which are frozen or affected by soft rot or wet breakdown.

(ii) U.S. No. 2 Potatoes for Processing: 15 percent for potatoes which fail to meet the requirements of this grade but not more than one-third of this tolerance or 5 percent for potatoes which are frozen or affected by soft rot or wet breakdown or are seriously damaged by internal defects, including therein not more than 1 percent for potatoes which are frozen or affected by soft rot or wet breakdown.

(2) For off-size:

(i) Undersize: 3 percent when the size specified is less than $2^1/_4$ inches in diameter or less than 6 ounces in weight; and, 5 percent when the size specified is $2^1/_4$ inches or more in diameter or 6 ounces or more in weight.

(ii) Oversize: 10 percent.

(3) For foreign material, other than sprouts, 2 percent.

(4) For sprouts, 2 percent.

(b) In the application of these standards to determine the

percentages of potatoes in any lot which meet the requirements of the respective grades or size categories no tolerances apply.

DEFINITIONS

§2851.3416 Similar Varietal Characteristics.

"Similar varietal characteristics" means that the potatoes in any lot have the same general shape, color and character of skin, and color of flesh. Round or intermediate types are not mixed with long types, red skinned with white skinned varieties, smooth skinned with russet varieties, and potatoes with yellowish flesh are not mixed with white-fleshed varieties.

§2851.3417 Fairly Well Shaped.

"Fairly well shaped" means that the potato is fairly regular in shape for the variety and is not materially affected by second growth, and is not materially pointed, dumbbell-shaped or otherwise materially ill-formed.

§2851.3418 Fairly Smooth.

"Fairly smooth" means that the potato is not deeply indented at the eyes or elsewhere, or materially ridged, folded or lumpy.

§2851.3419 Soft Rot or Wet Breakdown.

"Soft rot or wet breakdown" means any soft, mushy or leaky condition of the tissue such as slimy soft rot, leak, advanced jelly-end rot, wet type fusarium tuber rot, or wet breakdown following freezing injury or heat injury.

§2851.3420 Sprouts.

"Sprouts" means sprouts more than three-fourths inch in length which are attached to the tuber and detached sprouts of any length.

§2851.3421 Foreign Material.

"Foreign material" means stems, vines, adhering dirt, loose dirt, stones, trash and other extraneous substances other than sprouts.

§2851.3422 Damage.

"Damage" means any specific defect defined in this section, or an equally objectionable variation of any one of these defects, any other defect, or any combination of defects, which materially detracts from the processing quality of the potato, or which cannot be removed without a loss of more than 5 percent of the total weight of the potato, including peel covering the defective area, or which requires cuts in more than two places to remove the major defective portion in preparation for processing. The following specific defects are considered as damage:

(a) *Clipped ends*, when potatoes of a long variety have one or both ends clipped prior to trimming to remove defects at time of preparation for processing.

(b) *Shriveling,* when the potato is more than moderately shriveled, spongy, or flabby to the extent that it cannot be properly peeled in preparation for processing.

§2851.3423 Usable Piece.

"Usable piece" means the portion remaining after trimming a potato as it occurs in the sample, or at time of processing, which has more unusable material than permitted in U.S. No. 2 Potatoes for Processing: *Provided,* That the remaining portion meets the following requirements:

(a) Shall have not more than 10 percent unusable material;

(b) Meets the specified size requirements;

(c) In case of round or intermediate varieties represents not less than three-fourths of the untrimmed potato; and,

(d) In the case of long varieties:

(1) Weighs at least 4 ounces.

(2) If ends are clipped, cuts shall be approximately at right angles to the longitudinal axis of the portion of tuber.

(3) When one end is clipped, not more than an additional 25 percent of the surface may be smoothly trimmed in removing other defects, and the greatest diameter of the clipped end may not exceed the length of the remaining portion.

(4) When both ends are clipped, in addition to the area of the larger clipped end, not more than 25 percent of the surface may be smoothly trimmed in removing other defects, including the area of the smaller clipped end.

(5) When both ends are clipped, the length of the remaining portion shall be at least two-thirds of the combined greatest diameters of the two clipped ends.

(6) Slab or digger cuts shall be approximately parallel to the longitudinal axis of the potato and shall not expose more than 25 percent of the surface area of the potato.

§2851.3424 Serious Damage.

"Serious damage" means any defect, or any combination of defects, which seriously detracts from the processing quality of the potato, or which cannot be removed without a loss of more than 10 percent of the total weight of the potato, including peel covering the defective area.

Dated: June 4, 1963.
G. R. Grange,
Deputy Administrator,
Marketing Services

(F.R. Doc. 63-6016; Filed, June 6, 1963; 8:47 am)

APPENDIX 3

United States Standards for Grades of Canned White Potatoes

[Effective August 24, 1987]

This is the third issue of the United States Standards for Grades of Canned White Potatoes published in the *FEDERAL REGISTER* of July 23, 1987, (52 FR 27663) to become effective August 24, 1987. This issue supersedes the second issue of the United States Standards for Grades of Canned White Potatoes which have been in effect since December 1, 1977.

These standards are included in the Code of Federal Regulations, Title 7—Agriculture, Part 52.

Voluntary U.S. grade standards are issued under the authority of the Agricultural Marketing Act of 1946, which provides for the development of official U.S. Grades to designate different levels of quality. These grade standards are available for use by producers, suppliers, buyers, and consumers. As in the case of other standards for grades of processed fruits and vegetables, these standards are designed to facilitate orderly marketing by providing a convenient basis for buying and selling, for establishing quality control programs and for determining loan values.

The standards also serve as a basis for the inspection and grading of commodities by the Federal inspection service, the only activity authorized to approve the designation of U.S. grades as referenced in the standards, as provided under the Agricultural Marketing Act of 1946. This service, available as on-line (in-plant) or lot inspection and grading of all processed fruit and vegetable products, is offered to interested parties, upon application, on a fee-for-service basis. The verification of some specific recommendations, requirements, or tolerances contained in the standards can be accomplished only by the use of on-line inspection procedures. In all instances, a grade can be assigned based on final product factors or characteristics.

In addition to the U.S. grade standards, grading manuals or instructions for inspection of several processed fruits and vegetables are available upon request for a nominal fee. These manuals or instructions contain detailed interpretations of the grade standards and provide step-by-step procedures for grading the product.

Grade standards are issued by the Department after careful consideration of all data and views submitted, and the Department welcome suggestions which might aid in improving the standards in Future revisions. Comments may be submitted to, and copies of standards and grading manuals obtained from:

Chief, Processed Products Branch
Fruit and Vegetable Division, AMS
U.S. Department of Agriculture
P.O. Box 96456, Rm. 0709, So. Bldg.
Washington, DC 20090-6456

Sec.	
52.1811	Product description.
52.1812	Styles.
52.1813	Definitions of terms.
52.1814	Recommended fill of container.
52.1815	Recommended minimum drained weights.
52.1816	Recommended sample unit sizes.
52.1817	Size requirements for whole potatoes.
52.1818	Grades.
52.1819	Factors of quality.
52.1820	Requirements for grades.
52.1821	Determining the grade of a lot.

Authority: Agricultural Marketing Act of 1946, Secs. 203, 205, 60 Stat. 1087, as amended, 1090, as amended (7 U.S.C. 1622, 1624).

Note: Compliance with the provisions of these standards shall not excuse failure to comply with the provisions of the Federal Food, Drug, and Cosmetic Act, or with applicable State laws and regulations.

§52.1811 Product Description.

Canned white potatoes is the product as defined in the Standards of Identity for Certain Other Canned Vegetable (21 CFR 155.200) issued under the Federal Food, Drug, and Cosmetic Act.

§52.1812 Styles.

(a) Whole consists of peeled white potatoes that retain the approximate original conformation of the whole potato.

(b) Slices or sliced consist of peeled whole white potatoes cut into slices of substantially uniform thickness.

(c) Dice or diced consists of peeled whole white potatoes cut into cube-like units of substantially uniform size.

(d) Shoestring, french style, or julienne consists of peeled whole white potatoes cut into rectangular units having length measurements which are three (3) or more times the width measurements.

(e) Pieces consist of peeled whole white potatoes of random size and/or shape, or potatoes that have been cut into approximate quarters or wedge-shaped units.

(f) Any combination of two (2) or more of the foregoing styles constitutes a style and shall be considered as a mixture of the individual styles that comprise the combination.

§52.1813 Definitions of Terms.

As used in these U.S. standards, unless otherwise required by the context, the following terms shall be construed, respectively, to mean:

(a) Blemished means units affected by brown or black internal or external discoloration, discolored or unpeeled eyes, hollow heart, scab, or units blemished by other means to such an extent that the appearance or eating quality of the unit is materially affected.

(b) Seriously blemished means units affected by brown or black internal or external discoloration, pathological or insect injury or units blemished by other means to such an extent that the appearance or eating quality of the unit is seriously affected.

(c) Color.

(1) Good color means that the units, exclusive of blemished areas, are practically free from oxidation or light greenish coloration, and have a bright, practically uniform, light color, typical of canned white potatoes processed from potatoes of similar varietal characteristics.

(2) Reasonably good color means that the units possess a reasonably good color, and the units individually or collectively may be variable in color, dull, slightly oxidized, or otherwise discolored but not to the extent that the appearance of the product is seriously affected.

(3) Poor color means the units fail to meet the requirements for reasonably good color.

(d) Defects.

(1) Practically free from defects means the defects present do not materially affect the appearance or edibility of the product.

(2) Reasonably free from defects means the defects present do not seriously affect the appearance or edibility of the product.

(e) Diameter.

(1) ...of elongated whole potatoes means the greatest measurement at right angles to the longitudinal axis of the units.

(2) ...of round or nearly round whole potatoes means the greatest measurement across the center of the unit.

(3) ...of sliced style potatoes means the shortest measurement of the larger cut surface of the slice.

(f) Extraneous vegetable material (EVM) means harmless plant material such as leaves, stems, or roots.

(g) Flavor and odor.

(1) Good flavor and odor means a good, distinctive flavor and odor which is characteristic of properly prepared and properly processed canned white potatoes, (including any permitted safe and suitable optional ingredient (s)), that are free from objectionable flavors or odors.

(2) Reasonably good flavor and odor means that the canned white potatoes, (including any permitted safe and suitable optional ingredient (s)), may be lacking in good flavor and odor but are free from objectionable flavors and odors.

(h) Mechanical damage means damage incurred during harvesting or processing such as broken, crushed or cracked units, and units that are excessively trimmed.

(i) Pathological or insect injury means damage caused by disease or insects.

(j) Peel means the outer layer of the potato that is normally removed during processing.

(k) Potato unit means one whole, slice, dice, shoestring, or piece of potato as applicable for the style.

(l) Sample unit size means the amount of product to be used for grading. It may be:

(1) The entire contents of a container;

(2) A portion of the contents of a container;

(3) A combination of the contents of two (2) or more containers;

(4) A portion of unpacked product.

(m) Texture.

The factor of texture refers to the tenderness of the canned white potatoes and to the degree of freedom from sloughing and from hard or objectionably coarse grained units.

(1) Good texture means the texture of the potatoes is practically uniform and is typical of properly prepared and properly processed potatoes and that the potatoes are firm and tender and have a fine and even grain. There may be sloughing to a degree that does not more than slightly affect the appearance of the product.

(2) Reasonably good texture means the potatoes are reasonably tender, and may be variable in texture ranging from somewhat soft to firm, but are not tough, hard or mushy. There may be a moderate amount of sloughing that does not seriously affect the appearance of the product.

(3) Poor texture means the potato units fail to meet the requirements for reasonably good texture.

§52.1814 Recommended Fill of Container.

The fill of container is not incorporated in the grades of the finished product since fill of container, as such, is not a factor of quality for the purposes of these grades. Each container shall be filled with white potatoes as full as practicable without impairment of quality and the product and packing medium shall occupy not less than 90 percent of the total capacity of the container.

Total capacity of the container means the maximum weight of distilled water, at 68 degrees Fahrenheit (20 degrees Celsius), which the sealed container will hold.

§52.1815 Recommended Minimum Drained Weights.

(a) General.

(1) The minimum drained weight values are given in Table I. They are not incorporated in the grades of the finished product since

drained weight, as such, is not a factor of quality for the purposes of these grades.

(2) The minimum drained weights are based on the weight of the white potatoes after the canned product has been allowed to equalize for 15 days or more days after the product has been canned.

(b) Method of determining drained weight.

The drained weight of canned white potatoes is determined by emptying the contents of the container upon a U.S. Standard No. 8 circular sieve (or equivalent) of the proper diameter containing 8 meshes to the inch (0.0937-inch (2.4 mm), +/- 3 percent, square openings) so as to distribute the product evenly. Without shifting the product, incline the sieve to a 17 to 20 degree angle to facilitate drainage and allow to drain for two (2) minutes.

The drained weight is the weight of the sieve and white potatoes less the weight of the dry sieve. The diameter of the sieve shall be 8 inches (20.3 cm), or equivalent, if the water capacity of the container is less than 3 pounds (1.36 kg), or 12 inches (30.5 cm) or equivalent, if such capacity is 3 pounds (1.36 kg) or more.

(c) Compliance with minimum drained weight values.

Compliance with the minimum drained weight values in Table I is determined by averaging the drained weights from all the containers in the sample which represent a specific lot. Such lot is considered as meeting the minimum drained weight values if the following criteria are met:

(1) The sample average (average of all the containers in the sample) meets the minimum average drained weight value (designated as "XD" in Table I or IA); and

(2) The number of sample units which fail to meet the minimum drained weight value for individual containers (designated as "LL" in Table I or IA) does not exceed the acceptance number specified in the applicable sample plan for Lot Inspection and Certification of Processed Fruits and Vegetables and Related Products.

§52.1816 Recommended Sample Unit Sizes.

The requirements for size determination and for quality factors other than the defect defined as extraneous vegetable material are based on a recommended sample unit size of 567g (20.0 oz) of drained product, or the entire drained contents of a container. The recommended sample unit size for extraneous vegetable material is the entire contents of the container.

TABLE I — Minimum Drained Weights For Canned White Potatoes - [ENGLISH SYSTEM (AVOIRDUPOIS)]

Container Designation	Container Dimensions		Styles									
	Diameter (inches)	Height (inches)	Whole		Sliced		Diced		Juliene		Pieces	
			X_d	LL	X_d	LL	X_d	LL	X_d	LL	X_d	LL
			Ounces		Ounces		Ounces		Ounces		Ounces	
8z Tall	2 $^{11}/_{16}$	3-$^{4}/_{16}$	5.5	4.8	5.5	5.0	5.6	5.1	5.3	4.8	5.5	4.8
No. 300	3	4-$^{7}/_{16}$	9.5	8.7	9.7	9.0	10.0	9.5	8.8	8.3	9.5	8.7
No. 303	3-$^{3}/_{16}$	4-$^{6}/_{16}$	10.2	9.3	10.2	9.4	10.5	9.8	9.3	8.6	10.2	9.3
No.2	3-$^{7}/_{16}$	4-$^{9}/_{16}$	13.0	11.9	13.3	12.4	13.5	12.7	12.3	11.5	13.0	11.9
No.2-1/2	4-$^{1}/_{16}$	4-$^{11}/_{16}$	19.0	17.7	19.5	18.4	20.0	19.0	18.3	17.3	19.0	17.7
N0. 10	6-$^{3}/_{16}$	7	74.0	71.5	75.0	73.0	76.0	74.2	72.0	70.2	74.0	71.5

TABLE IA — Minimum Drained Weights For Canned White Potatoes - Metric System

Container Designation	Container Dimensions		Styles									
	Diameter	Height	Whole		Sliced		Diced		Juliene		Pieces	
	millimeters		X_d	LL	X_d	LL	X_d	LL	X_d	LL	X_d	LL
			grams		grams		grams		grams		grams	
8z Tall	68.3	82.6	155.9	136.1	155.9	141.7	158.8	144.6	150.3	136.1	155.9	136.1
No. 300	76.2	112.7	269.3	246.6	275.0	255.1	283.5	269.3	249.3	235.2	269.3	246.6
No. 303	81.0	111.1	289.2	263.7	289.2	266.5	297.7	277.8	263.7	243.8	289.2	263.7
No.2	87.3	115.9	368.5	337.4	377.0	351.5	382.7	360.0	348.7	326.0	368.5	337.4
No.2-1/2	103.2	119.1	538.6	501.8	552.8	521.6	567.0	538.6	518.8	490.4	538.6	501.8
N0. 10	157.2	177.8	2097.9	2027.0	2126.2	2069.5	2154.6	2103.5	2041.2	1990.1	2097.9	2027.0

§52.1817 Size Requirements for Whole Potatoes.

(a) A lot of canned whole potatoes shall be assigned a single size designation if the applicable requirements of Table II are met.

(b) A lot of canned whole potatoes that fails the requirements of Table II for a single size designation shall be declared in terms of individual sample unit size designations.

§52.1818 Grades.

(a) U.S. Grade A is the quality of canned white potatoes that meets the applicable requirements of Tables III through VII and scores not less than 90 points.

(b) U.S. Grade B is the quality of canned white potatoes that meets the applicable requirements of Tables III through VII and scores not less than 80 points.

(c) Substandard is the quality of canned white potatoes that fails to meet the requirements for U.S. Grade B.

§52.1819 Factors of Quality.

The grade of a lot of canned white potatoes is based on the following quality factors:

(a) Color;
(b) Uniformity of size and shape;
(c) Defects;
(d) Texture:
(e) Flavor and odor.

TABLE II — SIZE DESIGNATION OF WHOLE POTATOES

Word Designation	Number Designation	In 567g (20oz) Sample Unit:	One or More Sample Unit(s) May Fall In:
Tiny	Size 1	At least 453.6g (16oz) —80% are not more than 25mm (0.98in) in diameter. 113.4g (4oz)—20% may have a diameter of more than 25mm (0.98in) but not more than 38mm (1.49in).	The next larger designation—Small (size 2), provided these sample units do not exceed the acceptance number in Table IIA following.
Small	Size 2	At least 453.6g (16oz) —80% have a diameter of more than 25mm (0.98in) but not more 38mm (1.49in). 56.7g (2oz)—10% may have a diameter of 25mm (0.98in) or less. 56.7g (2oz)—10% may have a diameter more than 38mm (1.49in) but not more than 51mm (2in).	The next smaller designation—Tiny (size 1), the next larger designation—Medium (size 3) or a combination thereof, provided these sample units do not exceed the acceptance number in Table IIA following.
Medium	Size 3	At least 453.6g (16oz) —80% have a diameter of more than 38mm (1.498in) but not more 51mm (2in). 56.7g (2oz)—10% may have a diameter of more than 25mm (0.98in) but not more than 38mm (1.49in). 56.7g (2oz)—10% may have a diameter more than 51mm (2in).	The next smaller designation—Small (size 2), the next larger designation—Large (size 4) or a combination thereof, provided these sample units do not exceed the acceptance number in Table IIA following.
Large	Size 4	At least 510.3g (18oz) —90% have a diameter of more than 51mm (2in). 56.7g (2oz)—10% may have a diameter of more than 38mm (1.49in) but not more than 51mm (2in).	The next smaller designation—Medium (size 3), provided these sample units do not exceed the acceptance number in Table IIA following.

TABLE IIA

Number of sample units	3	6	13	21	29
Acceptance number	0	1	2	3	4

§52.1820 Requirements for Grades.

(a) **U.S. Grade A** is the quality of canned white potatoes that meets the applicable requirements of Tables III through VII and scores not less than 90 points.

(b) **U.S. Grade B** is the quality of canned white potatoes that meets the applicable requirements of Tables III through VII and scores not less than 80 points.

(c) **Substandard** is the quality of canned white potatoes that fails to meet the applicable requirements for U.S. Grade B.

§52.1819 Factors of Quality.

(a) Color;
(b) Uniformity of size and shape;
(c) Defects;
(d) Texture;
(e) Flavor and odor.

TABLE III — WHOLE STYLE

Quality Factors[1]	Grade A[2]	Grade B[3]
Flavor and odor:	Good	Reasonably
Color:	Good	Reasonably good
Score	18-20 points	16-17 points
Uniformity of size & shape:	Practically uniform	Reasonably uniform
In the 90% (by count) most uniform units, the weight of largest unit is not more than:	3.0 times the weight of the smallest unit	4.0 times the weight of the smallest unit
Score	18-20 points	16-17 points
Defects:	Practically free	Reasonably free
Mechanical damage, seriously blemished and blemished. Maximum:	113.4g (4.0 oz)[4]	170.1g (6.0 oz)[4]
Seriously blemished and blemished. Maximum:	56.7g (2.0 oz)[4]	113.4g (4.0 oz)[4]
Seriously blemished. Maximum:	28.4g (1.0 oz)[4]	56.7g (2.0 oz)[4]
Extraneous vegetable material (EVM)	1 piece/1.7 kg (60 oz) net wt. (sample avg.)	3 pieces/1.7 kg (60 oz) net wt. (sample avg.)
Sand, grit or silt.	None	Trace
Score	27-30 points	24-26 points
Texture:	Good	Reasonably good
Score	27-30 points	24-26 points
Total score	90-100 points	80-89 points

[1] All quality factors except EVM are based on a sample unit size of 567g (20 oz).

[2] Can be reasonably uniform in size and shape if total score is 90 points or more.

[3] Can fail requirements for reasonably uniform size and shape if total score is 80 points or more.

[4] Or 1 unit provided the lot average does not exceed the prescribed requirement.

TABLE IV — SLICED STYLE

Quality Factors[1]	Grade A[2]	Grade B[3]
Flavor and odor:	Good	Reasonably good
Color:	Good	Reasonably good
Score	18-20 points	16-17 points
Uniformity of size & shape:	Practically uniform	Reasonably uniform
Maximum thickness:	19 mm (0.74 in)	25 mm (0.98 in)
The diameter of the largest slice is not more than:	1.5 times the diameter of the second smallest slice	2.0 times the diameter of the second smallest slice
Score	18-20 points	16-17 points
Defects:	Practically free	Reasonably free
Mechanical damage, seriously blemished and blemished. Maximum:	85g (3.0 oz)	127.6g (4.5 oz)
Seriously blemished and blemished. Maximum:	56.7g (2.0 oz)	85g (3.0 oz)
Seriously blemished. Maximum:	14.2g (0.5 oz)	28.4g (1.0 oz)
Extraneous vegetable material (EVM)	1 piece/1.7 kg (60 oz) net wt. (sample avg.)	3 pieces/1.7 kg (60 oz) net wt. (sample avg.)
Sand, grit or silt.	None	Trace
Score	27-30 points	24-26 points
Texture:	Good	Reasonably good
Score	27-30 points	24-26 points
Total score	90-100 points	80-89 points

1 All quality factors except EVM are based on a sample unit size of 567g (20 oz).

2 Can be reasonably uniform in size and shape if total score is 90 points or more.

3 Can fail requirements for reasonably uniform size and shape if total score is 80 points or more.

TABLE V — DICED STYLE

Quality Factors[1]	Grade A[2]	Grade B[3]
Flavor and odor:	Good	Reasonably good
Color:	Good	Reasonably good
Score	18-20 points	16-17 points
Uniformity of size & shape:	Practically uniform	Reasonably uniform
Maximum allowance for irregular shaped units and units that are noticeably larger or smaller than the prevalent cube zize:	56.7g (2.0 oz)	141.8g (5.0 oz)
Score	18-20 points	16-17 points
Defects:	Practically free	Reasonably free
Mechanical damage, seriously blemished and blemished. Maximum:	56.7g (2.0 oz)	85g (3.0 oz)
Seriously blemished and blemished. Maximum:	22.7g (0.8 oz)	34g (1.2 oz)
Seriously blemished. Maximum:	5.7g (0.2 oz)	11.3g (0.4 oz)
Extraneous vegetable material (EVM)	1 piece/1.7 kg (60 oz) net wt. (sample avg.)	3 pieces/1.7 kg (60 oz) net wt. (sample avg.)
Sand, grit or silt.	None	Trace
Score	27-30 points	24-26 points
Texture:	Good	Reasonably good
Score	27-30 points	24-27 points
Total score	90-100 points	80-89 points

[1] All quality factors except EVM are based on a sample unit size of 567g (20 oz).

[2] Can be reasonably uniform in size and shape if total score is 90 points or more.

[3] Can fail requirements for reasonably uniform size and shape if total score is 80 points or more.

TABLE VI — FRENCH STYLE

Quality Factors[1]	Grade A[2]	Grade B[3]
Flavor and odor:	Good	Reasonably good
Color:	Good	Reasonably good
Score	18-20 points	16-17 points
Uniformity of size & shape:	Practically uniform	Reasonably uniform
Allowance for units less than 13 mm (0.51 in) in length:	56.7g (2.0 oz)	141.8g (5.0 oz)
Score	18-20 points	16-17 points
Defects:	Practically free	Reasonably free
Mechanical damage, seriously blemished and blemished. Maximum:	56.7g (2.0 oz)	85g (3.0 oz)
Seriously blemished and blemished. Maximum:	22.7g (0.8 oz)	34g (1.2 oz)
Seriously blemished. Maximum:	5.7g (0.2 oz)	11.3g (0.4 oz)
Extraneous vegetable material (EVM)	1 piece/1.7 kg (60 oz) net wt. (sample avg.)	3 pieces/1.7 kg (60 oz) net wt. (sample avg.)
Sand, grit or silt.	None	Trace
Score	27-30 points	24-26 points
Texture:	Good	Reasonably good
Score	27-30 points	24-26 points
Total score	90-100 points	80-89 points

1 All quality factors except EVM are based on a sample unit size of 567g (20 oz).

2 Can be reasonably uniform in size and shape if total score is 90 points or more.

3 Can fail requirements for reasonably uniform size and shape if total score is 80 points or more.

TABLE VII — PIECES

Quality Factors[1]	Grade A[2]	Grade B[3]
Flavor and odor:	Good	Reasonably good
Color:	Good	Reasonably good
Score	18-20 points	16-17 points
Uniformity of size & shape:	Practically uniform	Reasonably uniform
Maximum allowance for units weighing less than 7.1g (0.25 oz):	28.4g (1.0 oz)	56.7g (2.0 oz)
Of those units weighing 7.1g (0.25 oz) or more, the weight of the largest unit is not more than:	2.0 times the weight of the second smallest unit	4.0 times the weight of the second smallest unit
Score	18-20 points	16-17 points
Defects:	Practically free	Reasonably free
Mechanical damage, seriously blemished and blemished. Maximum:	85g (3.0 oz)	127.6g (4.5 oz)
Seriously blemished and blemished. Maximum:	56.7g (2.0 oz)	85g (3.0 oz)
Seriously blemished. Maximum:	14.2g (0.5 oz)	28.4g (1.0 oz)
Extraneous vegetable material (EVM)	1 piece/1.7 kg (60 oz) net wt. (sample avg.)	3 pieces/1.7 kg (60 oz) net wt. (sample avg.)
Sand, grit or silt.	None	Trace
Score	27-30 points	24-26 points
Texture:	Good	Reasonably good
Score	27-30 points	24-26 points
Total score	90-100 points	80-89 points

1 All quality factors except EVM are based on a sample unit size of 567g (20 oz).

2 Can be reasonably uniform in size and shape if total score is 90 points or more.

3 Can fail requirements for reasonably uniform size and shape if total score is 80 points or more.

§52.1821 Determining the Grade of a Lot.

The grade of a lot of canned white potatoes covered by these standards is determined by the procedures found in the "Regulations Governing Inspection and Certification of Processed Fruits and Vegetables, Processed Products Thereof, and Certain Other Processed Food Products" (7 CFR 52.1 through 52.83).

Washington, DC: July 16, 1987.

J. Patrick Boyle
Administrator

APPENDIX 4

United States Standards for Grades of Frozen French Fried Potatoes

[Effective February 8, 1967]

This is the third issue of the United States Standards for Grades of Frozen French Fried Potatoes. These standards are issued by the Department after careful consideration of all data and views submitted.

These standards were first published in the *Federal Register* of April 22, 1966 (31 FR 9190), to become effective July 1, 1966. They were amended as published in the Federal Register of January 24, 1967 (32 FR 779), effective February 8, 1967.

Issued under the authority of the Agricultural Marketing Act of 1946, which provides for the development of official U.S grades to designate different levels of quality, the grade standards are for the voluntary use of producers, suppliers, buyers, and consumers. As in the case of other standards for grades of processed fruits and vegetables, these standards are designated to facilitate orderly marketing by providing a convenient basis for establishing quality control programs, and for determining loan values.

These standards will also serve as a basis for the inspection and grading of this commodity Federal inspection service, which is also provided under the Agricultural Marketing Act of 1946. This service, available for inspection and grading of other processed products as well, is offered to interested parties, upon application, on a fee-for-service basis.

These standards are issued by the Department after careful consideration of all data and views submitted, and the Department welcomes suggestions which might aid in improving the standards in future revisions. Comments may be submitted to, and copies of standards obtained from:

Chief, Processed Products Branch
Fruit and Vegetable Division, AMS
U.S. Department of Agriculture
PO Box 96456, Rm. 0709, So. Bldg.
Washington, DC 20090-6456

Sec.

§52.2391	Product description.
§52.2392	Fry color type.
§52.2392a	Types.
§52.2393	Styles.
§52.2394	Length designations.
§52.2395	Grades.
§52.2396	Ascertaining the grade of a sample unit.
§52.2397	Ascertaining the rating for the factors which are scored.
§52.2398	Color.
§52.2399	Uniformity of size and symmetry.
§52.2400	Defects.
§52.2401	Texture.
§52.2402	Ascertaining the grade of a lot.
§52.2403	Ascertaining the length designation of a lot.
§52.2404	Equivalents.
§52.2405	Score sheet for frozen french fried potatoes.

Authority: The provisions of this subpart issued under Sec. 205, 60 Stat. 1090, as amended; 7 U.S.C. 1624.

Note: Compliance with the provisions of these standards shall not excuse failure to comply with the provisions of the Federal Food, Drug, and Cosmetic Act or with applicable State laws and regulations.

§52.2391 Product Description:

Frozen french fried potatoes are prepared from mature, sound, white or Irish potatoes (*Solanum tuberosum*). The potatoes are washed, sorted, and trimmed as necessary to assure a clean and wholesome product. The potatoes may or may not be cut into pieces. The potatoes are processed in accordance with good commercial practice which includes deep frying or blanching in a suitable fat or oil and which may include the addition of any ingredient permissible under the Federal Food, Drug, and Cosmetic Act. The prepared product is frozen and is stored at temperatures necessary for its preservation.

§52.2392 Fry Color Type.

(a) General: The term **fry color**, as used in this subpart, refers to the color change which occurs in the potato units solely because of the frying process. Potato units may be designated as to **fry color** in accordance with USDA color standards for frozen french fried potatoes. (See §52.2398 for further reference to these color standards.)

(b) The **fry color** of a sample unit may be designated in accordance with the following criterion:

Designation	Fry color of the units
Extra light	Lighter than USDA No. 1
Light	Similar to USDA No. 1 color
Medium light	Mostly similar to USDA No. 2. Predominantly lighter than No. 3 but may include No. 1.
Medium	Mostly similar to USDA No. 3. May include units of No. 4 and/or No. 2 fry color.
Dark	Predominantly darker than USDA No. 3. May include color No. 4 or darker.

§52.2392a Types.

(a) Frozen french fried potatoes are of two types, based principally on intended use as follows:

(1) *Retail type*. This type is intended for household consumption. It is normally packed in small packages which are labeled or marked for retail sales. It may be otherwise designated for such use.

(2) *Institutional type*. This type is intended for the hotel, restaurant, or other large feeding establishment trade. Primary containers, usually 5 pounds or more, are often not completely labeled as for retail sales.

(b) If it is not possible to ascertain the type, the quality requirements of the retail type apply.

§52.2393 Styles.

(a) *General*. The style of frozen french fried potatoes is identified by the general size, shape, or other physical characteristics of the potato units. Styles with cut units may be further identified by substyles as follows:

(1) *Strait* cut refers to smooth cut surfaces; and

(2) *Crinkle* cut refers to corrugated cut surfaces.

(b) *Strips*. This style consists of elongated pieces of potato with practically parallel sides and of any cross-sectional shape. This style may be further identified by the approximate dimensions of the cross-section, for example:

$^1/_4$ x $^1/_4$ inch
$^3/_8$ x $^3/_8$ inch
$^1/_2$ x $^1/_4$ inch

(1) *Shoestring* refers to strip, either straight cut or crinkle cut, with a cross-sectional are predominantly less than that of a square measuring $^3/_8$ x $^3/_8$ inch.

(c) *Slices*. This style consists of pieces of potato with two practically parallel sides, and which otherwise conform generally to the shape of the potatoes. This style may also contain a normal amount of outside slices.

(d) *Dices*. This style consists of pieces of potato cut into approximate cubes.

(e)*Rissole*. This style consists of whole or nearly whole potatoes.

(f) *Other*. Any other individually frozen french fried potato product may be designated as to style by description of the size, shape, or other characteristic which differentiates it from the other styles.

§52.2394 Length Designations.

(a) *General*. The length designations described in this section apply to **strip** styles only.

(b) *Criteria for length designations of a sample unit*. Frozen french fried potato strips are designated as to length in accordance with the following criteria. **Percent,** as used in this section, means the percent, by count, of all strips of potato that are $^1/_2$-inch in length, or longer.

(1) *Extra long*. Eighty (80) percent or more are 2 inches in length or longer, and 30 percent or more are 3 inches in length or longer.

(2) *Long*. Seventy (70) percent or more are 2 inches in length, or longer; and 15 percent or more are 3 inches in length or longer.

(3) *Medium*. Fifty (50) percent or more are 2 inches in length or longer.

(4) *Short*. Less than 50 percent are 2 inches in length or longer.

(c) Criteria for length designation of a lot. (See §52.2403 for these criteria.)

§52.2395 Grades.

(a) *U.S. Grade A* (or *U.S. Fancy)* is the quality of frozen french fried potatoes, except **strip** style, **short** length, that:

(1) Have a good flavor;

(2) Have a good color;

(3) Are practically uniform in size and symmetry;

(4) Are practically free from defects;

(5) Possess a good texture; and

(6) Score not less than 90 points when scored in accordance with the scoring system outline in this subpart.

(b) *U.S. grade A* (or *U.S. Fancy)* **Short** is the quality of frozen french fried potatoes of **strip** style, **short** length that:

(1) Have a good flavor;

(2) Have a good color;

(3) Are practically uniform in size and symmetry;

(4) Are practically free from defects:

(5) Possess a good texture; and

(6) Score not less than 90 points when scored in accordance with the scoring system outline in this subpart.

(c) *U.S. Grade B* (or *U.S. Standard*) is the quality of frozen french fried potatoes that:

(1) Have a reasonably good flavor;

(2) Have a reasonably good color;

(3) Are reasonably uniform in size and symmetry;

(4) Are reasonably free from defects;

(5) Possess a reasonably good texture; and

(6) Score not less than 80 points when scored in accordance with the scoring system outline in this subpart.

(d) *Substandard* is the grade of frozen french fried potatoes that fail to meet the requirements of **U.S. Grade B.**

§52.2396 Ascertaining the Grade of a Sample Unit.

(a) *General.* The grade of a sample unit of frozen french fried potatoes is ascertained by considering: the factor of flavor which is not scored; the ratings for the factors of color, uniformity of size and symmetry, defects, and texture which are scored; the total score; and the limiting rules which may be applicable.

(b) *Sample unit size.* For purposes of rating the quality factors, sample units are established as follows:

(1) *Retail type.* One pound of product (16 ounces) selected from a production line or from one or more market packages.

(2) *Institutional type.* Two pounds of product (32 ounces) selected from a production line or from one market package.

(c)*Definitions of flavor.*

(1) **Good flavor** means the good characteristic flavor and odor of properly prepared french fried potatoes. Such flavor is free from rancidity and bitterness, from pronounced scorched or caramelized flavors, and from off-flavors and off-odors of any kind.

(2) **Reasonably good flavor** means a flavor that may be somewhat lacking in good flavor and odor but is free from objectionable flavors and objectionable odors of any kind.

(d) *Heating the product.* When heating is required in connection with evaluating the quality factors, the sample unit is heated in accordance with the manufacturer's directions or in a manner that will give equivalent results. In the absence of such directions the following methods are prescribed:

(1) *Oven method.* This method is used for products packaged, labeled, marked, or otherwise designated to indicate probable retail sales for household use.

(i) Place the product while still in the frozen state on a piece of crumpled and partially straightened aluminum foil of sufficient size so that at least 9 ounces of the product may be spread in a single layer on the foil. The aluminum foil may be supported by a piece of sheet metal or a shallow pan.

(ii) Place foil and frozen contents into a properly ventilated oven preheated to 425°F and allow to remain 15 minutes or until the interior portions of the larger pieces are properly cooked.

(2) *Deep fat method.* This method is used for products which are prepared, packaged, labeled, or marked to indicate probable hotel, restaurant, or other feeding establishment use. Place the product in a suitable basket or rack and cook, by immersing in clean

suitable fat or oil at a temperature of about 365°F for $1^1/_2$ minutes or until the larger pieces are properly cooked.

(e) *Factors rated by score points.* The relative importance of each scoring factor is expressed numerically on the scale of 100. The maximum number of points that may be given each factor is:

Factors	Points
Color	30
Uniformity of size and symmetry	20
Defects	20
Texture	30
Total	100

§52.2397 Ascertaining the Rating for the Factors which are Scored.

The essential variations within each factor which is scored are so described that the value may be ascertained for each factor and expressed numerically.

The numerical range within each factor which is scored is inclusive. (For example: **18 to 20 points** means 18, 19, or 20 points.)

§52.2398 Color.

(a) *General.* The color of the product is evaluated before complete defrosting but after any exterior frost has evaporated. After heating, adjustment in the scores may be made to reflect color variations due to heating. The color numbers refer to a color which is visually similar to the designated color in the USDA Color Standards for Frozen French Fried Potatoes. The Fry Color designations in this section are the same as those described in §52.2392 Fry color type.

(b) *Availability of color standards.* The USDA Color Standard for frozen french fried potatoes cited in this sub-part are official color standards which may be used also to classify the fry color of other cooked products. Information about these color standards, and their availability, may be obtained from:

Chief, Processed Products Branch
Fruit and Vegetable Division, AMS
U.S. Department of Agriculture
PO Box 96456, Rm. 0709, So. Bldg.
Washington, DC 20090-6456

(c) *"A" classification*. Frozen french fried potatoes that have a good color may be given a score of 27 to 30 points. **Good color** means the bright characteristic color of properly prepared frozen french fried potatoes. The fry color may be **Extra light, Light, Medium light, or Medium** within the uniformity criterion in §52.2392. After heating, the product is practically free from units which vary markedly from the criterion for the fry color of the units.

(d) *"B" classification*. If the frozen french fried potatoes have a reasonably good color, a score of 24 to 26 points may be given. Frozen french fried potatoes that fall into this classification may not be graded above U.S. Grade B, regardless of the total score for the product (this is a limiting rule). **Reasonably good color** means a characteristic frozen french fry potato color which may be dull but not off color. The fry color may be variable, exceeding the uniformity criterion of **Extra light, Medium light, Medium or Dark.** After heating, however, the variation in the fry color of the units does not seriously affect the appearance of the product.

(e) *"SStd" classification*. Frozen french fried potatoes that fail to meet the requirements of paragraph (d) of this section may be given a score of 0 to 23 points and shall not be graded about substandard, regardless of the total score for the product (this is a limiting rule).

§52.2399 Uniformity of Size and Symmetry.

(a) *General*. In strip styles, the uniformity of length of normal shaped strips is not considered under this section. Such variation is considered under §52.2394. The following definitions apply to the specified styles:

(b) *Definitions*:

(1) *Chip*. In all styles except **Dices, chip** means a piece of potato that is less than one-half inch in its greatest direct dimension. A chip in Dices style is any piece of potato that is less than one half the volume of a normal dice. Chips are not considered small pieces for the purposes of this section.

(2) *Small piece*. A small piece in strip styles, means a strip that is less than 1 inch in length.

(3) *Sliver*. A sliver in strip styles is a strip of substantially smaller average cross-section than the predominant size of the strips and which has less than one-third of the weight of an average unit of the same length.

(4) *Irregular pieces.* Irregular pieces, in all styles, means pieces of potato that do not have the general conformation of the other units of the product.

(c) *"A" classification.* Frozen french fried potatoes that are practically uniform in size and symmetry may be given a score of 18 to 20 points. **Practically uniform in size and symmetry** has the following meanings for the styles specified:

(1) *Strip styles.*

(i) Any chips present no more than slightly detract from the appearance of the product; and

(ii) of all the product units, except chips not more than 15 percent, by count, may consist of small pieces, slivers, and/or irregular pieces.

(2) *All styles except strips.* Any chips present and/or any variation in the size and shape of the potato units may no more than slightly detract from the appearance of the products.

(d) *"B" classification.* If the frozen french fried potatoes are only reasonably uniform in size and symmetry a score of 16 or 17 points may be given. Frozen french fried potatoes that fall into this classification shall not be graded above U.S. Grade B, regardless of the total score for the product (this is a limiting rule). **Reasonably uniform in size and symmetry** has the following meanings for the styles specified:

(1) *Strip styles.*

(i) Any chips present may not seriously detract from the appearance of the product; and

(ii) Of all the product units, except **chips,** not more than 30 percent, by count, may consist of small pieces, slivers, and/or irregular pieces.

(2) *All styles except strip.* Any chips present and/or any variation in the size and shape of the potato units may not seriously detract from the appearance of the product.

(e) *"SStd" classification.* Frozen french fried potatoes that fail to meet the requirements of paragraph (d) of this section may be given a score of 0 to 15 points and shall not be graded above Substandard, regardless of the total score for the product (this is a limiting rule).

§52.2400 Defects.

(a) *General.* This factor is concerned with imperfections in the product, such as necrosis, crushed units, discolored eyes, callous areas, and discolorations which affect its appearance or edibility.

(b) *Insignificant imperfections.* Imperfections which individually are of little consequence and are not considered as defects for the purposes of these standards. Their effect, however, on the overall appearance or edibility of the product is considered. Among such insignificant imperfections are:

(1) Very light surface or internal discolorations of any size;

(2) Light brown surface or internal discolorations smaller in size that the area of a circle, or the volume of a sphere, of $^{3}/_{16}$ inch diameter; and

(3) Dark brown surface or internal discolorations smaller in size than the area of a circle, or the volume of a sphere, of $^{1}/_{8}$ inch diameter.

(c)*Minor defects.* Minor defects are imperfections which detract, but not seriously so, from the appearance or edibility of the product. Among such imperfections are:

(1) Light brown surface or internal discolorations equal to or larger in size than the area of a circle, or the volume of a sphere, of $^{3}/_{16}$ inch diameter but smaller in size than the area of a circle, or the volume of a sphere, of $^{5}/_{16}$ inch diameter; and

(2) dark brown surface or internal discolorations equal to or larger in size than the area of a circle, or the volume of a sphere, of $^{1}/_{8}$ inch diameter but smaller in size than the area of a circle, or the volume of a sphere of $^{1}/_{4}$ inch diameter.

(d) *Minor defective.* A minor defective is a unit of potato that is affected by one of more minor defects which detract, but not seriously so, from the appearance or edibility of the unit.

(e) *Major defects.* Major defects are imperfections of such a nature as to seriously detract from the appearance of edibility of the product. Among such imperfections are:

(1) Light brown surface or internal discolorations equal to or larger in size than the area of a circle, or the volume of a sphere, of $^{5}/_{15}$ inch diameter.

(2) dark brown surface or internal discolorations equal to or larger in size than the area of a circle, or the volume of a sphere, of $^{1}/_{4}$ inch diameter; and

(3) any condition of the potato which is offensive because of color, odor, character, or for any other reason.

(f) *Major defective.* A major defective is a unit of potato that is affected by one or more major defects or by any combination of major and/or minor defects so as to detract seriously from the appearance or edibility of the unit.

(g) *"A" classification.* Frozen french fried potatoes that are practically free from defects may be given a score of 18 to 20 points. **Practically free from defects** means that:

(1) Any combination of defects present (including insignificant imperfections) may no more than slightly detract from the appearance or edibility of the product;

(2) any carbon specks present may no more than slightly detract from the appearance of the product; and

(3) the minor and major defectives that may be present in the sample unit, do not exceed the allowances in Table I or Table II of this subpart.

(h) *"B" classification.* If the frozen french fried potatoes are reasonably free from defects a score of 16 to 17 points may be given. Frozen french fried potatoes that fall into this classification shall not be graded above U.S. Grade B, regardless of the total score for the product (this is limiting). **Reasonably free from defects** means:

(1) Any combination of defects present (including insignificant imperfection) does not seriously detract from the appearance or edibility of the product;

(2) any carbon specks present do not seriously detract from the appearance of the product; and

(3) the minor and major defectives that may be present in the sample unit do not exceed the allowances in Table I or Table II of this subpart.

(i) *"SStd" classification.* Frozen french fried potatoes that fail to meet the requirements of paragraph (h) of this section may be given a score of 0 to 15 points and shall not be graded above Substandard, regardless of the total score for the product (this is a limiting rule).

TABLE I — MAXIMUM ALLOWANCES FOR DEFECTIVES IN ALL STYLES EXCEPT SHOESTRING STYLE STRIPS AND DICES

Grade Classification	Type of Defects	Retail Type (Defectives in a 1-lb sample unit)	Institutional Type (Defectives in a 2-lb sample unit)
A and A short	Minor and major	5 defectives	18 defectives
	Limit for major	1 defectives	4 defectives
B	Minor and major	9 defectives	28 defectives
	Limit for major	2 defectives	8 defectives

TABLE II — MAXIMUM ALLOWANCES FOR DEFECTIVES IN SHOESTRING STYLE STRIPS AND DICES

Grade Classification	Type of Defects	Retail Type (Defectives in a 1-lb sample unit)	Institutional Type (Defectives in a 2-lb sample unit)
A and A short	Minor and major	9 defectives	28 defectives
	Limit for major	2 defectives	8 defectives
B	Minor and major	18 defectives	36 defectives
	Limit for major	5 defectives	12 defectives

§52.2401 Texture.

(a) *General.* The factor of texture is evaluated within 3 minutes after the product has been heated as specified, and while it is well above room temperature.

(b) *"A" classification.* Prepared french fried potatoes that possess a good texture may be given a score of 27 to 30 points. **Good texture** means that the external surfaces of the units moderately crisp, show no noticeable separation from the inner portion, and are not excessively oily; the interior portions are well cooked, tender, and practically free from sogginess; except that shoestring style strips and dices may be moderately crisp throughout.

(c)*"B" classification.* If the prepared french fried potatoes possess only a reasonably good texture a score of 24 to 26 points may be given. Frozen french fried potatoes that fall into this classification shall not be graded above U.S. Grade B regardless of the total score for the product (this is a limiting rule). **Reasonably good texture** means that the external surfaces of the units may be slightly hard or slightly tough, show no more than a moderate separation from the interior portion, and are not excessively oily; the interior portions are well cooked, reasonably tender, and reasonably free from sogginess.

(d) *"SStd" classification.* Frozen french fried potatoes that fail to meet the requirements of paragraph (c) of this section may be given a score of 0 to 23 points and shall not be graded above Substandard regardless of the total score for the product (this is a limiting rule).

§52.2402 Ascertaining the Grade of a Lot.

The grade of a lot of frozen french fried potatoes covered by these standards is determined by the procedures set forth in the Regulations Governing inspection and Certification of Processed Fruits and Vegetables. Processed Products Thereof, and Certain Processed Food Products (7 CFR 52.1 to 52.83).

§52.2403 Ascertaining the Length Designation of a Lot.

(a) This section applies to strip styles only.

(b) A lot of frozen french fried potato strips may be given a single length designation when the sample units representing the lot comply with the plan outlined in paragraph (c) of this section, in that:

(1) Any number of sample units, indicated by "X" are the designated length and/or longer length designation; and

(2) the number of sample units indicated by "D" does not exceed the acceptance number for the sample size in the sampling plan contained in this section.

(c) Plan for ascertaining the length designation of a lot:

Length Designation for the Lot	Individual Sample Units of Each Designation Permitted*			
	Extra Long	Long	Medium	Short
Extra long	X	D	None	None
Long	X	X	D	None
Medium	X	X	X	D
Short	D	X	X	X more than D

* Indicates limiting rule.

X means any number of sample units within the sample size.
D means the number of sample units indicated by the acceptance number for the sample size in the single sampling plan in this section.
X more than **D** means any number of sample units within the single sampling plan which is larger than the acceptance number for the sample size.

(d) Single sampling plan:

Sample size	6	13	21	29	38	48	60	72
Acceptance no.	1	2	3	4	5	6	7	8

§52.2404 Equivalents.

The following approximate equivalents may be used in connection with the numerical values used in this subpart:

1 inch	=	25.4 mm
2 inch	=	50.8 mm
3 inch	=	76.2 mm
$^1/_4$ inch	=	6.4 mm
$^1/_8$ inch	=	3.2 mm
$^3/_{16}$ inch	=	4.8 mm
$^5/_{16}$ inch	=	7.9 mm
1 pound	=	0.45 kg
425°F	=	218°C
365°F	=	185°C

§52.2405 Score sheet for frozen french fried potatoes.

Size and kind of container ______________________
Container mark of identification ________________
Label ____________________
Net weight (ounces) _________
Style and cross-sectional dimension _____________
Type ____________________
Length ___________________

Factors		Score Points	
Color	30	"A"	27-30
		"A" short	27-30
		"B"	24-26*
		"SStd"	0-23*
Uniformity of size & symmetry	20	"A"	18-20
		"A" short	18-20
		"B"	16-17*
		"SStd"	0-15*
Defects	20	"A"	18-20
		"A" short	18-20
		"B"	16-17*
		"SStd"	0-15*
Texture	30	"A"	27-30
		"A" short	27-30
		"B"	24-26*
		"SStd"	0-23*
Total score	100		
Flavor			
Grade			

* Indicates limiting rule.

Dated: February 8, 1997

/s/______________________
B.R. Grange, Deputy Administrator
Marketing Services

Amended effective Feb. 8, 1997
Published in *Federal Register*,
April 22, 1966 (31 FR 9190)
Amended January 24, 1967
(32 FR 779)

APPENDIX 5

United States Standards for Grades of Frozen Hash Brown Potatoes

[Effective November 15, 1976]

This is the first issue, of the United States Standards for Grades of Frozen Hash Brown Potatoes published in the *FEDERAL REGISTER* of October 1, 1976 (41 FR 43385) to become effective November 15, 1976.

Voluntary U.S. grade standards are issued under the authority of the Agricultural Marketing Act of 1946, which provides for the development of official U.S. grades to designate different levels of quality. These grade standards are available for use by producers, suppliers, buyers, and consumers. As in the case of other standards for grades of processed fruits and vegetables, these standards are designed to facilitate orderly marketing by providing a convenient basis for buying and selling, for establishing quality control programs, and for determining loan values.

The standards also serve as a basis for the inspection and grading of commodities by the Federal inspection service, the only activity authorized to approve the designation of U.S. grades as referenced in the standards, as provided under the Agricultural Marketing Act of 1946. This service, available as on-line (in-plant) or lot inspection and grading of all processed fruit and vegetable products, is offered to interested parties, upon application, on a fee-for-service basis. The verification of some specific recommendations, requirements, or tolerances contained in the standards can be accomplished only by the use of on-line inspection procedures. In all instances, a grade can be assigned based on final product factors or characteristics.

In addition to the U.S. grade standards, grading manuals or instructions for inspection of several processed fruits and vegetables are available upon request for a nominal fee. These manuals or instructions contain detailed interpretations of the grade standards and provide step-by-step procedures for grading the product.

Grade standards are issued by the Department after careful consideration of all data and views submitted, and the Department welcomes suggestions which might aid in improving the standards in future revisions. Comments may be submitted to, and copies of standards and grading manuals obtained from:

Chief, Processed Products Branch
Fruit and Vegetable Division, AMS
U.S. Department of Agriculture
PO Box 96456, Rm. 0709, So. Bldg.
Washington, DC 20090-6456

Section	
§52.6401	Product description
§52.6402	Styles
§52.6403	Grades
§52.6404	Determining the grade
§52.6405	Determining the rating for the factors which are scored
§52.6406	Color
§52.6407	Defects
§52.6408	Texture
§52.6409	Methods of analyses
§52.6410	Determining the grade of a lot
§52.6411	Score sheet for hash brown potatoes

Authority: Agricultural Marketing Act of 1946, Secs. 203, 205, 60 Stat. 1087, as amended, 1090, as amended (7 U.S.C. 1622, 1624).

Note: Compliance with the provisions of these standards shall not excuse failure to comply with the provisions of the Federal Food, Drug, and Cosmetic Act, or with applicable State laws and regulations.

§52.6401 Product Description.

Frozen hash brown potatoes are prepared from mature, sound, white or Irish potatoes (Solanum tuberosum) that are washed, peeled, sorted and trimmed to assure a clean and wholesome product. The potatoes so prepared are blanched, may or may not be fried and are shredded or diced or chopped and frozen and stored at temperatures necessary for their preservation.

§52.6402 Styles.

(a) *Shredded.* Shredded means potatoes cut into thin strips with cross-sectional dimensions from 1 mm by 2 mm to 4 mm by 6 mm and formed into a solid mass before freezing.

(b) *Diced.* Diced means potatoes cut into approximate cube shape units from 6 mm to 15 mm on an edge and loose frozen and which contain not more than 90 grams, per sample unit, of units smaller than one-half the volume of the predominant size unit.

(c) *Chopped.* Chopped means potatoes random cut into pieces predominantly less than 32 mm in their greatest dimension and loose frozen.

§52.6403 Grades.

(a) U.S. Grade A is the quality of frozen hash brown potatoes which has the following attributes:

(1) Normal flavor and odor;

(2) Good color;

(3) Practically free from defects;

(4) Good texture; and scores not less than 90 points when scored in accordance with the scoring system outlined in this subpart.

(b) U.S. Grade B is the quality of frozen hash brown potatoes which has at least the following attributes:

(1) Normal flavor and odor;

(2) Reasonably good color;

(3) Reasonably free from defects;

(4) Reasonably good texture; and scores not less than 80 points when scored in accordance with the scoring system outlined in this subpart.

(c) Substandard is the quality of frozen hash brown potatoes that fails to meet the requirements of Grade B.

§52.6404 Determining the Grade.

In addition to considering the other requirements outlined in this subpart, the following quality factors are evaluated:

(a) Factor not rated by score points. Flavor and odor.

(b) Factors rated by score points.

The relative importance of each factor which is scored is expressed numerically on a scale of 100. The maximum points that may be given such factors are:

Factors	Points
Color	30
Defects	40
Texture	30
Total	100

(c) *Sample unit size.* Sample unit size for evaluation of quality factors shall be 450 grams (15.9 ounces) for diced and chopped styles, and 340 grams (12 ounces) for shredded style.

(d) *Factor evaluation.*

(1) The factors of flavor and odor and color are determined before and after heating.

(2) The factor of defects is determined before heating.

(3) The factor of texture is determined after heating.

(e) *Normal flavor and odor* means the characteristic flavor and odor of properly prepared potatoes. Such flavor is free from bitterness, from pronounced scorched or caramelized flavors, and from off-odors of any kind.

§52.6405 Determining the Rating for the Factors Which are Scored.

The essential variations, within each factor which is scored, are so described that the value may be determined for each factor and expressed numerically. The numerical range within each factor, which is scored is inclusive (for example **27 to 30 points** means 27, 28, 29, or 30 points).

§52.6406 Color.

(a) *"A" classification.* Frozen hash brown potatoes that have a good color may be given a score of 27 to 30 points. **Good color** means the product, as a mass before heating, has a bright, uniform white to light cream color, which may include light brown units if processing included frying of any portion of the product. Such color is not more than slightly affected by yellowish units, greenish units, grayish units, or units which are otherwise discolored. After heating, the product surface shall have a fairly uniform brown color at least as dark as USDA No. 1 but not as dark as USDA No. 4 of the USDA Color Standards for Frozen French Fried Potatoes.

(b) *"B" classification*. Frozen hash brown potatoes that have a reasonably good color may be given a score of 24 to 26 points. Frozen hash brown potatoes that fall into this classification shall not be graded above U.S. Grade B, regardless of the total score for the product (this is a limiting rule). **Reasonably good color** means the product, as a mass before heating, has a reasonably bright, uniform white to cream color, which may include brown units if processing included frying of any portion of the product. Such color is not seriously affected by yellowish units, greenish units, grayish units, or units which are otherwise discolored. After heating, the product surface may brown only slightly or unevenly, but is not charred.

(c)*"SStd" classification*. Frozen hash brown potatoes that fail to meet the requirement of **"B"** classification may be given a score of 0 to 23 points and shall not be graded above substandard, regardless of the total score for the product (this is a limiting rule).

§52.6407 Defects.

(a) *General*. This factor is concerned with imperfections in the product, such as necrosis, crushed units, discolored eyes, and discolorations which affect its appearance or edibility.

(b) *Minor defects*. Minor defects are imperfections which detract only slightly from the appearance or edibility of the product. Among such minor defects are:

(1) Very light surface or internal discolorations of any size;

(2) Light brown surface or internal discolorations smaller in size than the area of a circle, or the volume of a sphere, of 4 mm diameter;

(3) Dark brown surface or internal discolorations smaller in size than the area of a circle, or the volume of a sphere, of 2 mm diameter; and

(4) In chopped and diced styles only, individual clusters of units which are not readily broken or separated manually, which weigh 5 grams or more but less than 10 grams.

(c) *Major defects.* Major defects are imperfections which materially detract from the appearance or edibility of the product. Among such major defects are:

(1) Light brown surface or internal discolorations equal to or larger in size than the area of a circle, or the volume of a sphere, of a 4 mm diameter but smaller in size than the area of a circle, or the volume of a sphere, of 8 mm diameter.

(2) Dark brown or internal discolorations equal to or larger in size than the area of a circle, or the volume of a sphere, of 2 mm diameter but smaller in size than the area of a circle, or the volume of a sphere, of 4 mm diameter; and

(3) In chopped and diced styles only, individual clusters of units which are not readily broken or separated manually, which weigh 10 grams or more.

(d) *Severe defects.* Severe defects are imperfections which seriously detract from the appearance or edibility of the product. Among such severe defects are:

(1) Light brown surface or internal discolorations equal to or larger in size than the area of a circle, or the volume of a sphere, of 8 mm diameter;

(2) Dark brown surface or internal discolorations equal to or larger in size than the area of a circle, or the volume of a sphere, of 4 mm diameter; and

(3) Any condition of a potato unit which is offensive because of color, odor, character, or for any other reason.

(e) *"A" classification.* Frozen hash brown potatoes that are practically free from defects may be given a score of 36 to 40 points. **Practically free from defects** means that:

(1) Any combination of defects present may no more than slightly detract from the appearance or edibility of the product; and

(2) The minor, major and severe defects, that may be present in the sample unit, do not exceed the allowance set forth in Table I.

(f) *"B" classification.* Frozen hash brown potatoes that are reasonably free from defects may be given a score of 32 to 35 points. Frozen hash brown potatoes that fall into this classification shall not be graded above U.S. Grade B, regardless of the total score for the product (this is a limiting rule). **Reasonably free from defects** means that:

(1) Any combination of defects present does not seriously detract from the appearance or edibility of the product; and

(2) The minor, major and severe defects; that may be present

in the sample unit, do not exceed the allowances set forth in Table I.

(g) *"SStd" classification.* Frozen hash brown potatoes that fail to meet the requirements of **"B"** classification, may be given a score of 0 to 31 points and shall not be graded above Substandard, regardless of the total score for the product (this is a limiting rule).

TABLE I — DEFECTS

Maximum Defect Allowance Per Sample Unit

(Chopped and Diced Styles: 450 g; Shredded Style: 340 g)

Defect	Grade A	Grade B
Chopped and Diced Styles		
Total Minor, Major and Severe	18	36
Total Major and Severe	9	18
Severe	2	4
Clusters 1/		
Total Minor and Major	6	18
Major	2	6
Shredded Style		
Total Minor, Major and Severe	14	27
Total Major and Severe	7	14
Severe	2	3

[1] The allowance for clusters in chopped and diced styles is in addition to the allowances for other defects that may be present.

§52.6408 Texture.

(a) The factor of texture is evaluated within three minutes after heating as specified in §52.6409 and while the product is well above room temperature.

(b) *"A" classification.* Prepared hash brown potatoes that have a good texture may be given a score of 27 to 30 points. **Good texture** means the potato units are firm and tender and the product is not more than slightly affected by units which are hard, mushy, pasty, soggy, dry or oil soaked.

(c) *"B" classification.* Prepared hash brown potatoes that have a reasonably good texture may be given a score of 24 to 26 points. Frozen hash brown potatoes that fall into this classification shall not be graded above U.S. Grade B, regardless of the total score for the product (this is a limiting rule). **Reasonably good texture** means the potato units are reasonably firm and tender and the product is not seriously affected by units which are had, mushy, pasty, soggy, dry, or oil soaked.

(d) *"SStd" classification*. Frozen hash brown potatoes that fail to meet the requirements of **"B"** classification may be given a score of 2 to 23 points and shall not be graded above Substandard, regardless of the total score for the product (this is a limiting rule).

§52.6409 Methods of Analyses.

Heating. The sample unit or a representative portion should be heated according to the manufacturers' directions, or in a manner which will give equivalent results.

In the absence of such directions the following methods may be used: Place approximately three tablespoons of cooking oil or fat in a frying pan—enough to properly fry the sample without excessive sticking—and preheat to 175°C (347°F). Add 170–225 grams (6–8 ounces) of the product and cook for 10 to 12 minutes; turning the product once or twice during cooking. A frying pan approximately 25 centimeters (10 inches) in diameter is appropriate for this method, however any suitable frying appliance may be used which will properly cook the desired amount of product.

§52.6410 Determining the Grade of a Lot.

The grade of a lot frozen hash brown potatoes covered by these standards is determined by the procedures set forth in the Regulations Governing Inspection and Certification of Processed Fruit and Vegetables, Processed Products Thereof, and Certain Other Processed Food Products (7 CFR 52.1 through 52.83).

§52.6411 Score Sheet for Hash Brown Potatoes.

Size and kind of container ______________________

Container mark of identification ________________

Label ____________________

Net contents ______________

Style ____________________

Color designation __________

Factors		Score Points	
Color	30	"A"	27-30
		"B"	24-26*
		"SStd"	0-23*
Defects	40	"A"	36-40
		"B"	32-35*
		"SStd"	0-31*
Texture	30	"A"	27-30
		"B"	24-26*
		"SStd"	0-23*
Total score	100		
U.S. grade			
Flavor and odor			

* Indicates limiting rule.

Effective date. This standard shall become effective November 15, 1976.

Dated: September 28, 1976.

Donald E. Wilkinson
Administrator
Agricultural Marketing Service

References and Further Reading

Agle, W. M. and G. W. Woodbury. 1968 Specific gravity–dry matter relationships and reducing sugar changes affected by potato variety, production area and storage. *Am Pot J* 45: 119–131.

Anon. 1987. *50 years: A foundation for the future*. Snack Food Association.

Baroudi, A. 1978. Evaluation of factors affecting color and acceptance of potato chips. Ph.D. Dissertation, The Ohio State University, Columbus, OH.

Brown, C. R. 1993. Origin and history of the potato. *Am Pot J* 70(5): 363–375.

Brown, J. B. and O. Smith. 1968 Effect of specific gravity of potatoes on sprouting in storage. *Am Pot J* 45: 265–266.

Cargill, B. F. 1976. *The Potato Storage*. American Society of Agricultural Engineers, St. Joseph, MI.

Cecil, I., A. Mackey, S. Joiner and J. Stockman. 1959. Homemakers' reactions to two specific gravity classes of Russett Burbank potatoes. *Am Pot J* 36: 81–89.

Chase, R. 1984. What to look for in new potato varieties. *Chipper Snacker* (February).

Chase, R. 1985. Increasing dry matter of potatoes producing optimum yields. *Chipper Snacker* (June), pp. 76–78.

Chubey, B. B. 1981. Sucrose rating proving a good measure at harvest. *Chipper Snacker* (February).

Cole, C. S. 1975. Variation in dry matter between and within potato tubers. *Potato Research* 18: 28–37.

Craft, C. C. and P. H. Heinze. 1951. Association of specific gravity with weight of individual tubers in late crop potatoes. *Am Pot J* 28: 580–582.

Cunningham, C. E., H. J. Murphy, M. J. Goven and R. V. Akeley. 1959. Yields, specific gravity and maturity of potatoes. *Maine Agr Exp Stat Bull* 379 (February).

Cunningham, C. E. and F. J. Stevenson. 1963. Inheritance of factors affecting potato chip color and their association with specific gravity. *Am Pot J* 40: 253–264.

Davis, R. E., Jr. 1962. Tissue airspace in the potato: Its estimation and relation to dry matter and specific gravity. *Am Pot J* 39: 298–305.

Deppen, J. 1968. The effect of potato varieties and storage temperatures upon the rates of accumulation of monosaccharides, disaccharides, organic acids and amino acids. Ph.D. Dissertation, The Ohio State University, Columbus, OH.

Edgar, A. D. 1951. Determining the specific gravity of individual potatoes. *Am Pot J* 28: 729–731.

Fitzpatrick, T. J., W. L. Porter and G. V. C. Houghland. 1969. Continued studies of the relationship of specific gravity to total solids of potatoes. *Am Pot J* 46: 120–127.

Fong, S. F. and E. S. Redshaw. 1973 A simple device for the determination of potato tuber specific gravity. *Am Pot J* 50: 254–256.

Gould, W. A. 1957. Density can be a 'tool' for control of quality. *Food Packer* (January).

Gould, W. A. and G. Clark. 1963. Potato chip research plant. *Farm & Home Res* 48(6): 88–89.

Gould, W. A. and J. Deppen. 1969. Handling potatoes for chipping. *Ohio Report* 54(4): 54–55.

Gould, W. A. 1977. Examining the potato; Part I. *Chipper / Snacker* 34: 26–28.

Gould, W. A. 1978. *Progress report on potato chip snack food research activities*. European Potato Chip Assn. (June).

Gould, W. A., B. Hair and A. Baroudi. 1979. Evaluation of potato cultivars before and after storage regimes for chipping. *Am Pot J* 56: 133–144.

Gould, W. A. 1983. Color evaluation of potato chips. *Chipper / Snacker* (September).

Gould. W. A. 1984. Evaluating new potato cultivars for chip manufacture. *Chipper / Snacker* (March).

Gould, W. A. 1985. Changes and trends in the snack food industry. *Cereal Foods World* 30(3): 219–220.

Gould, W. A. 1985. *Quality assurance manual for the manufacture of potato chips and snack foods*. Potato Chip/Snack Food Assn., Alexandria, VA.

Gould, W. A. 1985. Snack food quality evaluation. *Chipper / Snacker* (July).

Gould, W. A. and S. Plimpton. 1985. Quality evaluation of potato cultivars for processing. *North Central Regional Res Bull* 305 (August).

Gould, W. A. 1986. Procurement of chipping potatoes. *Chipper / Snacker* (January).
Gould, W. A. 1988. Quality of potatoes for chip manufacture. Potato Assn. of America Symposium, pp. 10–21.
Gould, W. A. 1989. Specific gravity—Its measurement and use. *Chipping Potato Handbook*, Snack Food Assn., Alexandria, VA.
Gould, W. A. 1994. *Snack food manufacture and quality assurance manual* (April). Snack Food Assn., Alexandria, VA.
Goulston, A. B. 1964. Solvent extraction technique for removing excess surface oil. U.S. Patent No. 3,127,271, March.
Greenwood, M. L., M. H. McKendrick and A. Hawkins. 1952. The relationship of the specific gravity of six varieties of potatoes to their mealiness as assessed by sensory methods. *Am Pot J* 29: 192-196.
Habib, A. T. and H. D. Brown. 1956. Factors influencing the color of potato chips. *Food Tech* X(7): 332–336.
Hannigan, J. 1981. Fat replacer cuts calories in fatty foods. *Food Eng* 53(13): 8–9.
Hasagawa, S., R. M. Johnson and W. A. Gould. 1966. Effect of cold storage on chlorogenic acid content of potatoes. *Agric & Food Chem* 14(2): 165–169.
Hasen, J. C. 1970. Storage and handling of potatoes for the chip industry. *Potato Chip Inst Int'l* (August).
Haynes, K. G., R. E. Webb, R. W. Goth and D. R. Wilson. 1989. The correlation of yield and specific gravity in the USDA potato breeding program. *Am Pot J* 66: 587–592.
Heinz, P. H., C. C. Craft, B. M. Mountjoy, and M. E. Kirkpatrick. 1952. Variations in specific gravity of potatoes. *Am Pot J* 29: 31–37.
Hill, M. K. and W. A. Gould. 1970. Effect of storage conditions on chip quality of potatoes. *J Food Sci* 42(4): 927–930.
Hooker, W. J. 1981. Compendium of potato diseases. APS Press.
Hope, G. W. 1964. A monograph for calculating specific gravity of potato tubers. 41: 221–222.
Houghland, G. V. C. 1966. New conversion table for specifc gravity, dry matter and starch in potatoes, 43: 138.
Hughes, J. C. 1954. Testing of hydrometers. *U.S. Dept. of Commerce, NBS Circ* 555.
Ifenkwe, O. P., E. J. Allen and D. C. E. Wurr. 1974. Factors affecting the relationship between tuber size and dry matter content. *Am Pot J* 51: 233–241.

Iritani, W. M., W. C. Sparks and W. H. Weinheimer. 1964. Factors that affect accuracy determining specific gravity. *Idaho Potato and Onion News,* Vol. III, No. 7.
Iritani, W. M. and L. D. Weller. 1976. Relationship of specific gravity to sugar accumulation in stored Norgold and Russet Burbank potatoes. *Am Pot J* 53: 57–65.
Johansen, R. H., J. C. Miller, D. W. Newsom and J. F. Fontenot. 1967. The influence of environment on the specific gravity, plant maturity and vigor of potato progenies. *Am Pot J* 44: 107–121.
Johnson, R. N. 1957. Factors affecting the yield, fat absorption and color of potato chips. Ph.D. Dissertation, The Ohio State University, Columbus, OH.
Joiner, S. and A. Mackey. 1962 Weight loss, specific gravity and mealiness during storage of Russet Burbank potatoes. *Am Pot J* 39: 320–325.
Killick, R. J. and A. W. Macarthur. 1980. The relationship between bruising and specific gravity in some potato varieties. *Pot Res* 23: 457–461.
Kleinkopf, G. E., D. T. Westermann, M. J. Wille and G. D. Kleinschmidt. 1987. Specific gravity of Russet Burbank potatoes. *Am Pot J* 64: 579–587.
Kleinschmidt, G. 1984. Laws of specific gravity: Keys to high profits. *University of Idaho Potato Specialist Report.*
Kunkel, R., J. Gregory and A. M. Binkley. 1951. Mechanical separation of potatoes into specific gravity shows promise for the potato chip industry. *Am Pot J* 28: 690–696.
Kunkel, R. and N. Holstad. 1972. Potato chip color, specific gravity and of potatoes with N-P-K. *Am Pot J* 49: 43–62.
Kushman, L. J. and F. L. Haynes, Jr. 1971. Influence of intercellular space differences due to variety and storage upon tuber specific gravity-dry matter relationships. *Am Pot J* 48: 173–181.
Lana, E. P. Potato production in North Dakota. 1976. *N. D. Ext Bull* 26.
Lana, E. P., R. H. Johansen and D. C. Nelson. 1970. Variation in specific gravity of potato tubers. *Am Pot J* 47: 9–12.
Lulai, E. C. 1986. Potato specific gravity. *Chipper/Snacker* (August).
Lulai, E. C. and P. H. Orr. 1979. Influence of potato specific gravity on yield and oil content of chips. *Am Pot J* 56: 379–390.
Lujan, L. and O. Smith. 1964. Potato quality XXV. Specific gravity and after cooking darkening of Kathadin potatoes as influenced by fertilizers. *Am Pot J* 41: 274–278.

Lyman, S. and A. Mackey. 1961. Effect of specific gravity, storage, and conditioning on potato chip color. *Am Pot J* 38: 51–57.
Maas, E. F. 1971. Effect of specific gravity and seed maturity on potato yields in peat soil. *Am Pot J* 48: 113–115.
Mackey, A., S. Joiner, I. Cecil and J. Stockman. 1958. Survey of specific gravity of Russet Burbank potatoes grown in Central Oregon and the Klamath Basin, 1957 crop. *Am Pot J* 35: 739–741.
Mancuso, J. and A. Caposeela. 1968. Low fat potato chips. U.S. Patent No. 3,402,049, September.
Murphy, Hugh J. and Michael J. Goven. 1959. Factors affecting the specific gravity of the white potato in Maine. *Maine Agr Exp Stat Bull* 583.
Murphy, H. J., M. J. Goven, and D. C. Merriam. 1966. Effect of three viruses on yield, specific gravity, and chip color of potatoes in Maine. *Am Pot J* 43: 393–396.
Motes, J. E. and J. K. Greig. 1970. Specific gravity, potato chip color and tuber mineral content as affected by soil moisture and harvest dates. *Am Pot J* 47: 413–418.
Nack, H. and H. G. Shutz. 1965. Oil spray for dry potato products. U.S. Patent No. 3,205,074, September.
Ng, K. C., H. D. Brown, R. H. Blackmore and J. Bushnell. 1957. The relation of the calcium content of potato tubers to the quality of potato chips. *Food Tech XI*(2): 118–122.
Nissen, M. 1955. The weight of potatoes in water. *Am Pot J* 32: 332–339.
Nonaka, M., M. L. Weaver and R. N. Sayre. 1974. Reducing the surface oil content of fried foods. *Food Tech* 28(5): 50–65.
Nylund, R. E. and A. J. Poivan. 1953. The influence of variety and date of planting on the relative cooking quality of potatoes graded according to specific gravity. *Am Pot J* 30: 107–118.
O'Keefe, R. B. 1979. New Potato varieties. *Chipper/Snacker* (February).
O'Keefe, R. B. 1980. Performance of new chipping potato varieties and selections. *Chipper/Snacker* (April).
Pearsall, L. W. 1952. Facts about potato quality. *Am Pot J* 29: 182–185.
Plimpton, S. 1985. Factors affecting the oil content of potato chips and the application of near-red reflectance to on-line moisture and oil content measurements. Ph.D. Dissertation, The Ohio State University, Columbus, OH.
Porter, W. L., T. J. Fitzpatrick and E. A. Talley. 1964. Studies of the relationship of specific gravity to total solids of potatoes. *Am Pot J* 41: 329–336.

Quarmby, A. R. 1981. Determining the radial distribution of specific gravity within potato tubers. *J Food Sci* 46: 508–514.
Rasmussen, M. P. 1955. Potato specific gravity studies. *Am Pot J* 32: 192–194.
Rastovski, A., A. van Es, et al. 1981. *Storage of potatoes*. Center for Agricultural Publishing and Documentation, Wageningen, Germany.
Ratcliffe, J. D. 1975. A practical analysis of the in-plant peeling losses on potatoes. *Food Trade Rev* (October).
Redshaw, E. S. and S. F. Fong. 1972. A specific gravity calculator. *Am Pot J* 49: 250–351.
Reeve, R. A., M. L. Weaver and H. Timm. 1971. Anatomy and compositional variations within potatoes. IV. Total solids distribution in different cultivars. *Am Pot J* 48: 269–277.
Rose, D. H. and H. T. Cook. 1949. Handling, storage, transporation and utilization of potatoes. *USDA Bibliographical Bull* 11 (December).
Rowe, R. 1977. Potato late blight. NPC/SFA Chipping Potato Seminar, March.
Saini, G. R. 1964. Determination of dry matter in potatoes. *Am Pot J* 40: 123–125.
Sawyer, R. L. 1979. The future role of the potato in the world. *New Zealand Potato Bull* (June).
Sayre, R. N., M. Nonaka and M. L. Weaver. 1975. French fry quality related to specific gravity and solids content variation among potato strips within the same tuber. *Am Pot J* 52:.73–82.
Schark, A. E., C. E. Peterson, and F. Carlin. 1956. The influence of variety on the specific gravity–mealiness relationship of potatoes. *Am Pot J* 33: 79–83.
Schippers, P. A. 1976. The relationship between specific gravity and percentage of dry matter in potato tubers. *Am Pot J* 543: 111–112.
Schoenemann, J. 1983. Producing tubers with high dry matter. *Am Veg Grower* (June), pp. 36–37.
Schoenemann, J. 1987. Ensure adequate dry matter. *Am Veg Grower* (December), pp. 36–39.
Schwartz, J. H., R. B. Greenspun and W. L. Porter. 1966. Chemical composition of potatoes V. Further studies on the relationship of organic acid concentrations to specific gravity and storage time. *Am Pot J* 43: 361–366.
Shaper, L. A., E. C. Yeager and P. H. Orr. 1986. Potato pile pressures. *Chipper / Snacker* (August).

Sharma, M. K., D. R. Isleib and S. T. Dexter. 1958. Specific gravity of different zones within potato tubers. *Am Pot J* 35: 784–788.

Shoemaker, J. S. 1947. *Vegetable Growing*. John Wiley & Sons, New York, NY.

Sijbring, P. H. and J. van der Velde. 1969. Principles of vacuum frying and the results of vacuum frying of chips in practice. *Food Trade Rev* 39(6): 39–42.

Silva, G. H. and W. T. Andrew. 1983. Sprouting of potato tubers in relation to specific gravity. *Am Pot J* 60: 563–565.

Smith, O. 1951. Potato quality. *Am Pot J* 28: 732–737.

Smith, O. 1960. NPCI Research Report. *Potato Chipper* (September), pp. 48–50.

Smith, O. 1962. Effect of source and rate of potash application on yields and specific gravity of potatoes and their processing quality. *Potato Chipper* (January), pp. 92–94.

Smith, O. 1977. *Potatoes: production, storing, processing*. The AVI Publishing Co., Inc., Westport, CT.

Sowokinos, J. R. 1980. Three applications of sucrose-rating to minimize sugar accumulations in stressed potatoes. *Chipper / Snacker* (September), pp. 29–30.

Tai, G. C. C., G. C. Misener, E. S. Allaby and L. P. McMillan. 1985. Grav-O-Tater: A computer apparatus for measuring specific gravity. *Am Pot J* 62: 403–408.

Talburt, W. F. and O. Smith. 1975. *Potato processing*. The AVI Publishing Co., Inc., Westport, CT.

Terman, G. L., M. Goven, and C. E. Cunningham. 1950. Effect of storage temperature and size on French fry quality, shrinkage and specific gravity of Maine potatoes. *Am Pot J* 27: 417–424.

Thornton, R. E. 1986. Rapid detection of potato tuber bruise damage. *Chipper / Snacker* (July).

Thornton, E. E., and J. B. Sieczka. 1980. Commercial potato production in North America. *Potato Association of America Handbook* (Supplement), *Am Pot J* 57.

Timm, H. and F. G. Merkle. 1963. The influence of chlorides on yield and specific gravity of potatoes. *Am Pot J* 40: 1-8.

Unrau, A. M. and R. E. Nyland. 1957. The relationship of physical properties and chemical composition to the mealiness in the potato: I. Physcial properties. *Am Pot J* 34: 245–253.

Unrau, A. M. and R. E. Nyland. 1957. The relationship of physical properties and chemical composition to the mealiness in the potato: II. Chemical composition. *Am Pot J* 34: 303–311.

Vanasse, N. A., I. D. Jones and H. L. Lucas. 1951. Specific gravity–dry matter relationship in potatoes. *Am Pot J* 28: 781–791.

Van der Schild, J. H. W. 1977. *Industrial processing of potato products*. Institute for Storage and Processing Agricultural Products (October).

Von Schelle, C., G. Svensson and J. Rasmusson. 1937. Die Bestimmung des Starkegehalts und der Trockensubstanz der Kartoffel mit Hilfe des spezifischen Gewichits. *Landw Vers Sta* 127: 67–96.

Willard, M. 1993. Potato Processing: Past, present and future. *Am Pot J* 70(5): 405–418.

Williams, A. E. 1951. *Potato Crisps*. Food Trade Press, Ltd., London, England.

Wilson, J. H. and A. M. Linsay. 1969. The relationship between specific gravity and dry matter content of potato tubers. *Am Pot J* 46: 323–328.

Wittenberger, R. T. 1951. Changes in specific gravity, starch content, and sloughing of potatoes during storage. *Am Pot J* 28: 738–747.

Woodbury, G. W. and W. H. Weinheimer. 1965. Specific gravity–solids correlations in Russet Burbank with respect to point of origin and storage history. *Am Pot J* 42: 98–104.

Young, D. A., P. W. Voisey and N. Dixon. 1964. A specific gravity calculator for potatoes. *Am Pot J* 41: 401–405.

Yungen, J. A., A. S. Hunter, and T. H. Bond. 1958. The influence of fertilizer treatments on the yield, grade, and specific gravity of potatoes in Eastern Oregon. *Am Pot J* 35: 386–395.

Zaehringer, M. V., R. M. Reeve, E. A. Talley, H. H. Murphy, D. H. Dinkle and R. B. Hyde. 1965. The estimation of total solids from specific gravity measurements. *Potato Handbook*, pp. 46–48.

Zaehringer, M. V., R. M. Reeve, E. A. Talley, D. H. Dinkle and R. B. Hyde. 1966. Specific gravity and composition of potatoes and various processing and cooking purposes. *Potato Handbook*, pp. 5–10.

CONTENTS INDEX

FIGURES INDEX

FSC
www.fsc.org
MIX
Papier | Fördert gute Waldnutzung
FSC® C083411

Kolibri 360 GmbH
Ferdinand-Jühlke-Straße 7
99095 Erfurt, Deutschland
produktsicherheit@kolibri360.de